Deutsches Zentrum für Entwicklungstechnologien – GATE

Deutsches Zentrum für Entwicklungstechnologien – GATE – stands for German Appropriate Technology Exchange. It was founded in 1978 as a special division of the Deutsche Gesellschaft für Technische Zusammenarbeit (GTZ) GmbH. GATE is a centre for the dissemination and promotion of appropriate technologies for developing countries. GATE defines „Appropriate technologies" as those which are suitable and acceptable in the light of economic, social and cultural criteria. They should contribute to socio-economic development whilst ensuring optimal utilization of resources and minimal detriment to the environment. Depending on the case at hand a traditional, intermediate or highly-developed can be the „appropriate" one. GATE focusses its work on four key areas:

– *Technology Exchange*: Collecting, processing and disseminating information on technologies appropriate to the needs of the developing countries; ascertaining the technological requirements of Third World countries; support in the form of personnel, material and equipment to promote the development and adaptation of technologies for developing countries.

– *Research and Development*: Conducting and/or promoting research and development work in appropriate technologies.

– *Cooperation in Technological Development:* Cooperation in the form of joint projects with relevant institutions in developing countries and in the Federal Republic of Germany.

– *Environmental Protection:* The growing importance of ecology and environmental protection require better coordination and harmonization of projects. In order to tackle these tasks more effectively, a coordination center was set up within GATE in 1985.

GATE has entered into cooperation agreements with a number of technology centres in Third World countries.

GATE offers a free information service on appropriate technologies for all public and private development institutions in developing countries, dealing with the development, adaptation, introduction and application of technologies.

Deutsche Gesellschaft für Technische Zusammenarbeit (GTZ) GmbH

The government-owned GTZ operates in the field of Technical Cooperation. 2200 German experts are working together with partners from about 100 countries of Africa, Asia and Latin America in projects covering practically every sector of agriculture, forestry, economic development, social services and institutional and material infrastructure. – The GTZ is commissioned to do this work both by the Government of the Federal Republic of Germany and by other government or semi-government authorities.

The GTZ activities encompass:

– appraisal, technical planning, control and supervision of technical cooperation projects commissioned by the Government of the Federal Republic or by other authorities

– providing an advisory service to other agencies also working on development projects

– the recruitment, selection, briefing, assignment, administration of expert personnel and their welfare and technical backstopping during their period of assignment

– provision of materials and equipment for projects, planning work, selection, purchasing and shipment to the developing countries

– management of all financial obligations to the partner-country.

Deutsches Zentrum für Entwicklungstechnologien – GATE
in: Deutsche Gesellschaft für Technische Zusammenarbeit (GTZ) GmbH
Postbox 51 80
D-6236 Eschborn 1
Federal Republic of Germany
Tel.: (0 61 96) 79-0   Telex: 41 523-0 gtz d

Albrecht Kaupp

Gasification of Rice Hulls

Theory and Praxis

A Publication of
Deutsches Zentrum für Entwicklungstechnologien – GATE
in: Deutsche Gesellschaft für Technische Zusammenarbeit (GTZ) GmbH

Springer Fachmedien Wiesbaden GmbH

The Author:
Albrecht Kaupp, PhD, staff member of GTZ/GATE has been working in the fields of civil engineering, mathematics, and biomass energy conversion systems since 1972. Now project officer for biomass energy conversion systems since 1983. His field of expertise is gasification of biomass.

CIP-Kurztitelaufnahme der Deutschen Bibliothek

Kaupp, Albrecht:
Gasification of rice hulls : theory and praxis ;
a publ. of Dt. Zentrum für Entwicklungstechno-
logien – GATE in: Dt. Ges. für Techn. Zusammen-
arbeit (GTZ) GmbH / Albrecht Kaupp. –
Braunschweig ; Wiesbaden : Vieweg, 1984.
ISBN 978-3-528-02002-6 ISBN 978-3-322-96308-6 (eBook)
DOI 10.1007/978-3-322-96308-6

Ursprünglich erschienen bei Friedr. Vieweg & Sohn Velagsgesellschaft mbH, Braunschweig 1984

A word of warning!

This book about gasification of rice hulls is based on fundamental research and a comparison of various systems. Because the gasification of rice hulls is very sensitive to scaling up or down a gasifier, many but not all of the physical and chemical properties of rice hulls are described in detail.

To obtain a convincing decision on the most suitable way of gasification of rice hulls the principle of negative selection was applied, starting with conventional woodgas producers. In order to demonstrate the various systems and show their disadvantages they are discussed in detail at the end of this book.

However, it must be said that, based on a great deal of experimental work only two ways of gasification of rice hulls seem to be technically feasible and promising in the requested range of 20-130 kWel.

1. An open core downdraft gas producer with no throat and a slowly rotating grate for ash removal. The limiting factor for this system is certainly the diameter of the reactor which should not be below 40 cm due to the caking of rice hulls.

2. Gasification of rice hulls in pelletized form. Although the densification of rice hulls is difficult and expensive, it allows their gasification in smaller reactors due to a more stable fuel-bed.

Option 1 is in the authors opinion the simplest and most economical way to gasify rice hulls although a great deal of work remains in order to predict the kinetics of rice hull gasification and optimize all the fine details which make the difference between a good and bad gasifier.

Eschborn, February 16, 1984 Albrecht Kaupp

ACKNOWLEDGEMENT

This study has received much technical, financial and moral support from many people and institutions. I would like to thank Professor John R. Goss for his patience. Eldon Beagle, Consultant, for the many hours he spent in talking with me about his vast experience in the utilization of rice hulls. Professor Stephan Whitaker, Department of Chemical Engineering, for his concise, uncompromising teaching and treatment of the theory of reactor design which was a keystone to most of the theoretical parts of this research.

My thanks are extended to the Carl Duisburg Gesellschaft in West Germany which financed part of this research. The Weyerhaeuser Company and their two-year scholarship which allowed me to work independently. Special thanks go to the Briggs and Stratton Corporation, Doug Janish and Mr. Robert Catterson, whose generous financial support over two years made this project possible.

The help of Bart Duff, International Rice Research Institute, Philippines, to secure adequate funding was greatly appreciated.

I thank George Giannini and the workshop people for the many gas producers and other devices they built for me. Also not forgotten is Kurt Creamer, Graduate Student, who assisted me in many of the experiments and patiently corrected my English with weekly awards for the "worst sentence". The illustrations for this work have been done by Jim Bumgarner. The many hundreds of pages were typed over and over again by Karin Clawson.

My greatest appreciation goes to Filiz who hates rice hulls, but certainly does understand the author.

CONTENTS

LIST OF FIGURES

PAGE

PAGE

PAGE

PAGE

PAGE

PAGE

LIST OF TABLES

I. INTRODUCTION

Man discovered fire and with it the combustion process between 0.5 and 1.4 million years ago. It took him much longer, until 1669, to develop the art of generating a combustible gas from a carbonaceous fuel (Wyer, 1907). It may be appropriate to ask why it took so long. It is even more appropriate to explore why the history of gasification was so chaotic and why the heyday of gasification was so short. Over 12,000 large gasifiers were installed in North America within a period of 30 years, providing electricity and street lights to many large cities. Moreover, in the 1930's and 1940's over 1 million automotive gas producers were operated worldwide. The period of commercial use of gas producers was quite short, considering the almost total disappearance of the technology after 1950.

The history of gasification is so astonishing and instructive, that it will be dealt with in a separate chapter.

Insofar as writing a dissertation is a serious matter, the time and efforts invested in this task should not be wasted by duplicating previous work or by trying to demonstrate banalities such as a discussion of the generation of a combustible gas from a carbonaceous fuel. An extensive survey in 63 countries has preceded this work to provide a basis for establishing research priorities in this field. (Kaupp and Goss 1982).

While conducting this survey it became clear that the level of disagreement among researchers, engineers and manufacturers is unusually high. The disagreement begins with the definition of the word "gasification". Of course, the mistake lies in the attempt to characterize a process with an imprecise tool, i.e. language. The literature is full of catch words such as fast pyrolysis, slow pyrolysis, pyrolysis without oxygen, pyrolysis with oxygen, starved air combustion, combustion with excess air, combustion in an oxidizing atmosphere, combustion in a reducing atmosphere and of course gasification itself. Naturally, finding a meaning for gasification within the mentioned framework is a hopeless undertaking. It is therefore appropriate to start at the beginning and make the statement:

> "The carbon in the fuel burns above a certain temperature when in contact with oxygen and this process generates a phenomena which we sense as heat. In addition, the process generates certain products, the natures and quantities of which are open for discussion".

It is a simple calculation to determine the amount of air necessary to completely burn the solid carbon in the fuel with the assumption that there are only three products: unburned residues other than carbon, water (H_2O) and carbon dioxide (CO_2). Whether it is possible to control the process such that those are the only three components is irrelevant for this definition. This process is known as stoichiometric combustion or combustion with 100% theoretical air, and is thus uniquely defined. The difficulties arise when no restrictions are imposed on the products and when less than 100% theoretical air is assumed. There are many definitions of the meaning of gasification, therefore there is certainly no danger in postulating still one more which hopefully is more precise.

One can demonstrate a combustion process taking place at less than stoichiometric air. Such a process will be described as occurring at an equivalence ratio ϕ which is defined as:

$$\phi = \frac{\text{Mass of Reaction Combustion Air}}{\text{Mass of Stoichiometric Air for Reaction}}$$

By definition ϕ is a number larger than 0, although in most combustion reactors values between 0.1 and 4.0 are achieved. Products from combustion processes are more than just solid residues and gaseous products. An appreciable amount of condensable liquids are also generated which may or may not be combustible. In converting reactants to products the chemical processes are endothermic or exothermic. However, energy can be stored in only two modes in the products:

a. Sensible heat which represents translational, vibrational and rotational energy of the molecules.

b. Chemical energy which upon further oxidation can be converted into sensible heat.

It is therefore rather intriguing to characterize gasification as the process which stores the maximum of usable chemical energy in the gaseous portion of the products. Although at first glance this seems another weak definition of gasification, it has the advantage of clearly laying out the objectives. A demonstration is given in Figure 1-1 which illustrates how the total energy stored in the gaseous phase changes with ϕ. One can also deduce another general concept which holds true for any combustion process: The total energy in the process is shifted into the gaseous phase with increased ϕ. In theory one can therefore utilize the ultimate elemental analysis of a biomass fuel and calculate the equivalence ratio ϕ which optimizes the process with respect to the chemical energy stored in the gaseous phase.

One objective of this work is to determine whether operation at this optimal ϕ can be sustained, and if so to determine the design that will accomplish this. In order to show the complexity of the problem, one need only list the major issues involved:

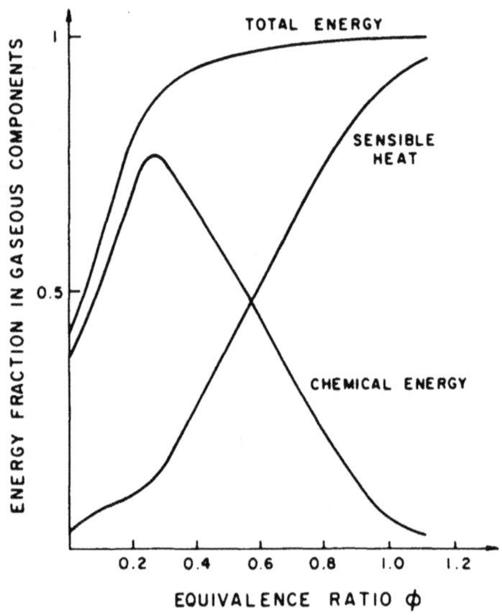

Figure 1-1. Energy fractions in gaseous components as a function of the equivalence ratio ϕ (Reed, 1979).

1. A gas producer by itself is of little use. In any installation a gasifier is either used to:
   a. generate a combustible gas which is subsequently burned in a burner to provide heat, or
   b. generate a gas which is used as a fossil fuel substitute in internal combustion engines. In this case there is a wide range of applications such as generation of electricity and mechanical power.

2. Almost all combustion processes generate a very complex mixture of gaseous, liquid and solid products. Some of these products are not only harmful to engines and burners but also to man. There is the tendency to minimize this aspect of gasification on a small scale because it requires some effort to detect and quantify these undesirable products. In order to illustrate this point a partial list of components of a combustion process is given in Table 1-1.

3. Depending on the usage of the producer gas very different restrictions are imposed on its composition. It is not clear whether the process of thermal decomposition at $\phi < 1$ can in all instances satisfy these requirements. On the other hand, it is well known as to what extent the various applications can utilize solid and condensable products.

   a. A center-fired burner can cope with solid and condensable contaminants in the gaseous phase much better than an internal combustion engine. The degree to which solid carbon particles and liquid higher hydrocarbons are burned off is however a strong function of the H/C and C/O ratios. Where the heat is used for drying purposes, it is important to point out that direct contact of the hot combustion products of producer gas with the material to be dried is desirable from an efficiency standpoint. However it is not known whether combustion of producer gas completely destroys undesirable contaminates which could otherwise affect the product to be dried.

   b. Extensive experience with producer gas in internal combustion

Table 1-1. Products of Thermal Decomposition of Biomass (Bugge, 1927)

Carbon monoxide	Formic acid	Methacrylic acid
Carbon dioxide	Acetic acid	γ-Butyrolactone
Hydrogen	Propionic acid	n-Butyric acid
	Crotonic acid	iso-Butyric acid
Water	iso-Crotonic acid	Angelic acid
Tiglic acid	Δ^3-Hexenone-2	2-Methylfurane
Δ^2-Pentenoic acid	Methyl n-butyl ketone	3-Methylfurane
γ-Valerolactone	3,6-Octanedione	Dimethylfurane
n-Valeric acid	2-Acetylfurane	2,5-Dimethyltetrahydrofurane
Methylethylacetic acid	Cyclopentanone	
n-Caproic acid	2-Methyl- Δ^2 cyclopentenone	Trimethylfurane
		5-Ethyl-2-methyl-4,5-dihydrofurane
Isocaproic acid	Methylcyclopentenolone	
n-Heptoic acid	Cyclohexanone	Coumarone
Lignoceric acid	Methylcyclohexenone	Pyroxanthone
Furoic acid	Dimethylcyclohexenone	Benzene
Methyl alcohol		Toluene
Ethyl alcohol	Phenol	Isopropylbenzene
Allyl alcohol	o-, m, and p-Cresol	m-Xylene
Propyl alcohol	o-Ethylphenol	Cymene
Methylvinylcarbinol	2,4-Dimethylphenol	Naphthalene
Isobutyl alcohol	3,5-Dimethylphenol	1,2,4,5-Tetramethylbenzene
Isoamyl alcohol	Catechol	
	Guaiacol	Chrysene
Formaldehyde	2-Methoxy-4-methylphenol	
Acetaldehyde		Ammonia
Propionaldehyde	2-Methoxy-4-vinylphenol	Methylamine
Valeraldehyde	2-Methoxy-4-ethylphenol	Dimethylamine
Isovaleraldehlyde	2-Methoxy-4-propylphenol	Trimethylamine
Trimethylacetaldehyde		Pyridine
Furfural	1,2-Dimethoxy-4-methylbenzene	3-Methylpyridine
5-Methylfurfural		Dimethylpyridine
Hydroxymethylfurfural	2,6-Dimethoxyphenol	
	2,6-Dimethoxy-4-methylphenol	Methane
Methylal		Heptadecane
Dimethylacetal	2,6-Dimethoxy-4-propylphenol	Octadecane
Acetone	Propylpyrogallol monomethyl ether	Eicosane
Methyl ethyl ketone		Heneicosane
Diacetyl	Coerolignol (or - on)	Docosane
Methyl propyl ketone	Euppittonic acid (or eupitton)	Tricosane
Methyl isopropyl ketone		Furane
Diethyl ketone		
Ethyl propyl ketone		

engines has clearly shown that piston engines cannot operate on producer gas which is contaminated over a specified limit with solid carbon and condensable products. Restrictions on the quality of the gas are severe. After cleaning, a practical limit exists of 5 - 15 mg of solids and tar per kg of gas if the gas is used as an engine fuel. A further constraint in using producer gas as an internal combustion engine fuel is its combustion characteristics, such as ignition characteristics, flame speed and soot formation. Thus, the amount of contaminants in the producer gas is not the only parameter that affects its suitability as an engine fuel. The actual composition of the gas will also have a pronounced effect on engine performance.

4. It is of necessity that a gas producer have an effective gas cleaning train when using the gas for internal combustion engines. The difficulties in cleaning the gas are rarely documented by scientific data. Some of the issues are:

   a. Purification of producer gas is not an impossible task and in principle not significantly different from the gas purification that has been performed for a century on a commerical scale. The technology and knowledge is available.

   b. Purification of producer gas results in most cases in bulky purification systems that often cause high pressure drops across the system. It is therefore important to find an acceptable compromise between purification efficiency, pressure drop and installation size.

c. In many applications the internal combustion engine is used as a suction pump. It draws the necessary combustion air into the gas producer and then draws the products of gasification through the purification train. A high pressure drop across the gas producer purification train is therefore highly undesirable for a suction gas producer, because it reduces the volumetric efficiency of the engine.

5. There are many arguments against using a gas producer to provide fuel for an internal combustion engine that operates under a variable load such as an automotive engine. The two most important are:

   a. A gas producer and gas cleaning train replaces only the gasoline tank of an engine. For automotive applications, there are substantial space and weight limitations with regard to the unit.

   b. A gas producer can be optimized for only one gas output rate and will usually not operate as well at a different rate. In particular, the formation of large amounts of condensable products are often a problem at non-optimal flow rates when fuels high in volatiles are used.

6. For any engine application it is desirable to have a gas composition that does not fluctuate greatly over time for the following reasons:

   a. Some components of the gas, such as hydrogen (H_2), can greatly influence the flame speed of the producer gas-air mixture, which in turn would make variable ignition timing as a function of the H_2 content necessary. See Figure 1-2.

b. The ever present moisture in the gas also affects the flame speed. Moisture also influences the reversible carbon (soot) reactions.

c. A changing heating value of the gas, which is entirely a function of its composition, will change the power output of the engine.

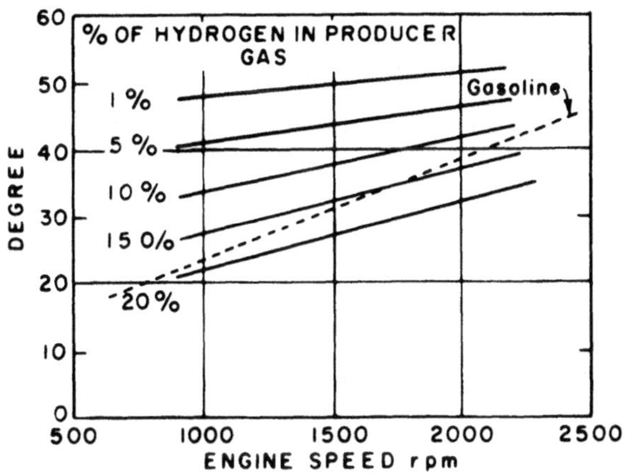

Figure 1-2. Ignition advancement (degrees before top dead center) versus hydrogen content of producer gas (Anonymous).

7. A gas producer is too often thought of as a simple device that can generate a combustible gas out of any biomass fuel. This misconception has had and will most likely continue to have a considerable impact on any future developments of the technology.

Having outlined the general issues it is worthwhile to refine the above points and to discuss the design parameters which are associated with these points.

1.1 Using a gas producer to generate a combustible gas may not have the advantages one would hope for with regard to fossil fuel savings, economics and ease of operation.

a. Generation of producer gas and subsequent combustion in a burner is no different from direct combustion of the biomass in a furnace with respect to the heat released. In both cases the product of combustion is a gas having a high sensible heat content. The alleged economic difference lies in the costly purification of the stack gases from a furnace operation. Whether gasification has an advantage over combustion in a furnace if the only objective is to generate heat for drying or space heating depends on many conditions.

b. Using producer gas as a substitute for fossil fuel in internal combustion engines is more attractive than direct combustion of producer gas in that the end product is a higher, more versatile form of energy: mechanical power and electricity. On the other Hand it also imposes more stringent requirements on the quality of the gas such as:

Low solid particle content

Low condensable content

High atomic H/C ratio

Steady gas composition

High H_2 content (see Figure 1-2)

Producer gas induction temperature near that of the ambient air.

Of special interest is the high H/C ratio. Most producer gas mixtures are in an unfavorable range of 0.1-1.4 for their atomic H/C ratio. The lower number represents a typical updraft producer gas very high in CO with almost no H_2 while the higher ratio of 1.4 is common for downdraft producer gas.

Soot formation can be expected for stoichiometric mixtures of producer gas and air as shown in Figures 1-3 and 1-4. It is of course always possible to supress soot formation by combustion at $\phi > 1$, as is done in burners. However, combustion at $\phi > 1$ in an internal combustion engine leads to a drastic decrease in the power output of the engine as shown in Figure 1-5. One can therefore expect soot formation within the combustion chamber at an optimal power output. This sensitivity of the power output with respect to ϕ is not as severe for an engine operated on gasoline.

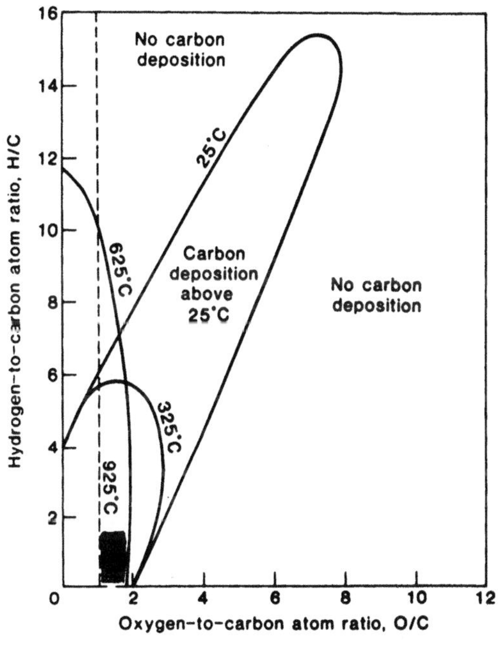

Figure 1-3. Soot formation as a function of H/C and O/C ratio (Probstein, 1982) (■ range for producer gas)

2.1 It is often useful to determine the amount of volatile matter given up when heating a carbonaceous fuel at low temperatures in an oxygen-free atmosphere. This pyrolytic process is what takes place in the

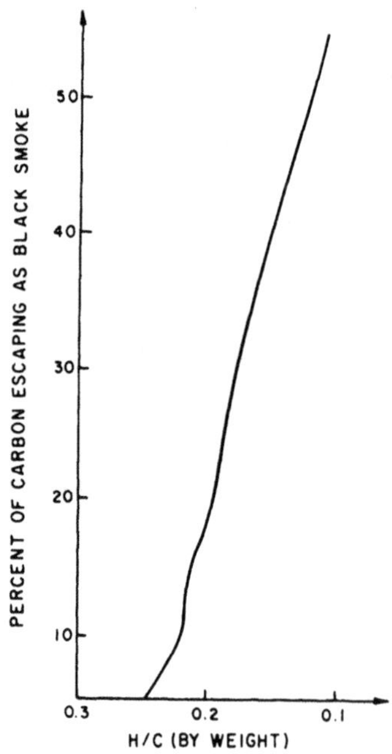

Figure 1-4. Soot formation as a function of H/C ratio (R. Reed, 1981)

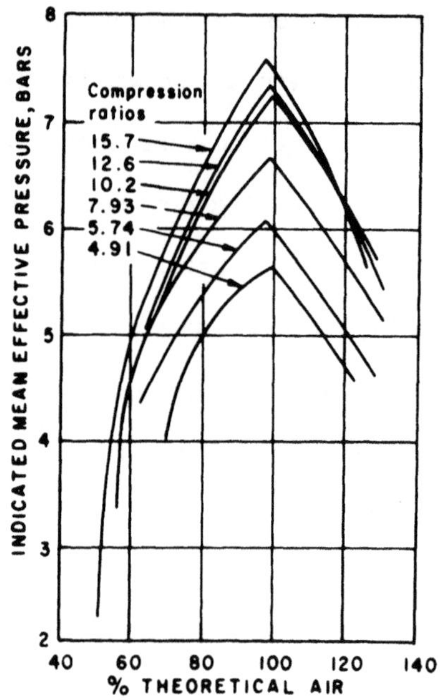

Figure 1-5. Power output as a function of ϕ (Kennedy, 1940)

upper zones of a gas producer where no oxygen is present and temperatures are around 100°C - 500°C. Of interest are:

a) The fraction of solid, liquid, and gaseous products formed during thermal decomposition at low temperatures.

b) The amount of carbon evolved and stored in the various phases.

c) The composition of the gaseous portion.

3.1 In most gasifiers the proportion of condensable products generated is substantial. However, this does not mean that these products must leave the gas producer in a condensable form, which will necessitate removal by a gas cleaning train for engine use. A properly designed gas producer includes provisions for the conversion of a major portion of the condensable products into noncondensable gases. However it is not easy to accomplish good tar conversion. The benefits of converting tarry products inside the gasifier are obvious:

a) Tar is a high energy component of the products of gasification having a higher heating value of 20-34 $MJkg^{-1}$ depending on its water content. It is therefore more economical to break it down into a mixture of combustible gases inside the gas producer, instead of removing it mechanically outside the gasifier.

b) The removal of tar vapors from the gas stream is costly, difficult and leads to bulky gas cleaning equipment.

c) Tar vapors and liquids have been known to be a health hazard for decades (Harvey, 1982).

Although the mechanism of tar conversion is difficult to explain, one may in general postulate three major conversion paths:

d) Thermal cracking at high temperatures which results in cracking of large hydrocarbon molecules.

e) Thermochemical conversion, which involves not only high temperatures but also a catalyst to promote cracking.

f) Combustion of the tar with oxygen.

Because of cost considerations, the catalytic effect needed for thermochemical conversion should come from the biomass fuel itself, rather than adding a catalyst just to support tar cracking (see Chapter 9).

4.1 One of the decisions that must be made with regard to gas cleaning equipment is whether to use wet or dry cleaning devices. This decision is mainly based on the relative amount of solid and liquid contaminants in the gas and not so much on their actual make-up. Some of the parameters involved are:

a) Solid particle load (mg kg^{-1} of raw gas)

b) Determination of the proportion of solid particles in the raw gas which can be attributed to physical entrainment as opposed to those which were generated by reversible carbon reactions (soot formation).

c) Amount of condensable products in the raw gas, their distillation point curves and the water content.

d) Dew point of the mixture, which influences greatly the location of any dry gas cleaning apparatus, since condensation of the vapors would render it useless.

5.1 Gas production in a gasifier is proportional to the rate of carbon gasified. It is therefore rather difficult to provide an internal combustion engine with the necessary gas quality if the engine is frequently changing speeds. Of special interest in this context are:

a) The reactivity of the fuel. It will determine the response time of the gas producer to a changing gas demand.

b) The ability of the gas producer to maintain the same vertical temperature profile under varying conditions. This characteristic will determine the amount of unconverted tar leaving a downdraft gasifier.

c) The ability of a gasifier to operate under a wide range of gas production rates at similar equivalence ratios.

6.1 Maintaining a steady gas composition and a suitable gas quality is dependent on two key parameters:

a) Existence of an undisturbed, homogeneous fuel bed, that is free of channels and caves.

b) Steady fuel flow and residue removal.

7.1 It is unrealistic to expect a gasifier to operate satisfactorily on various types of fuels. One hundred years of gasification research and commercial applications have clearly shown that the key to successful gasification is a gasifier specially designed for a particular fuel. It is of paramount importance that the physical and chemical characteristics of the fuel do not change significantly. The parameters that influence fixed bed gasification of a biomass fuel are listed with respect to their importance:

a) Physical shape of fuel particles and fuel bed structure

b) Moisture content

c) Volatile matter

d) Ash content

e) Ash composition

f) Energy content.

A large variety of biomass fuels have very similar chemical compositions on an ash and moisture-free basis, as shown in Figure 1-6. Therefore, the differences are attributed to local variations on how the fuel is processed, where it is grown, and to what climate it is exposed.

Summary

The foregoing discussion on gasification of biomass has emphasized the complexity of the process which does not only generate a combustible gas but other products as well. How many factors are involved in the process is shown in Figure 1-7. In the past the following issues seem to have dominated gas producer-engine applications:

a) Contaminations such as tar and dust in producer gas.

b) Development of compact, light and efficient gas cleaning trains.

c) Design of gas producers with a good turn down ratio.

d) Design of gas producers which could gasify inferior fuels such as bituminous coal, high ash fuels and fuels with low bulk densities.

e) Fuel preparations such as moisture content, fuel size, shape and fine particle content.

f) Safety and health hazards of producer gas operations.

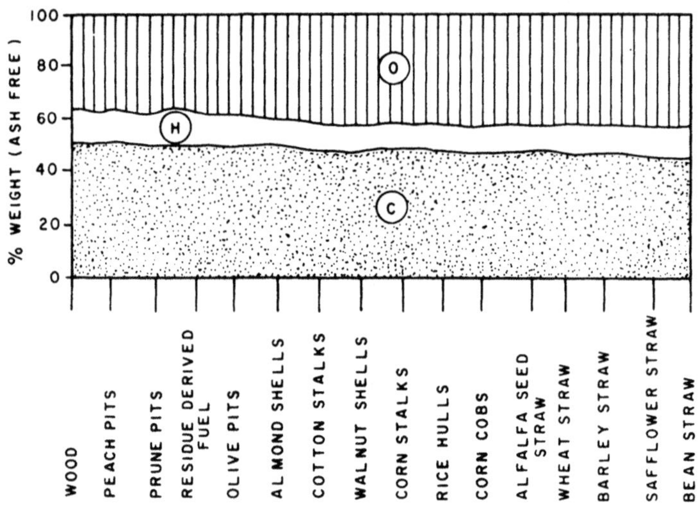

Figure 1-6. Ultimate elemental analysis on an ash and moisture free basis of various biomass fuels (Kaupp, 1982)

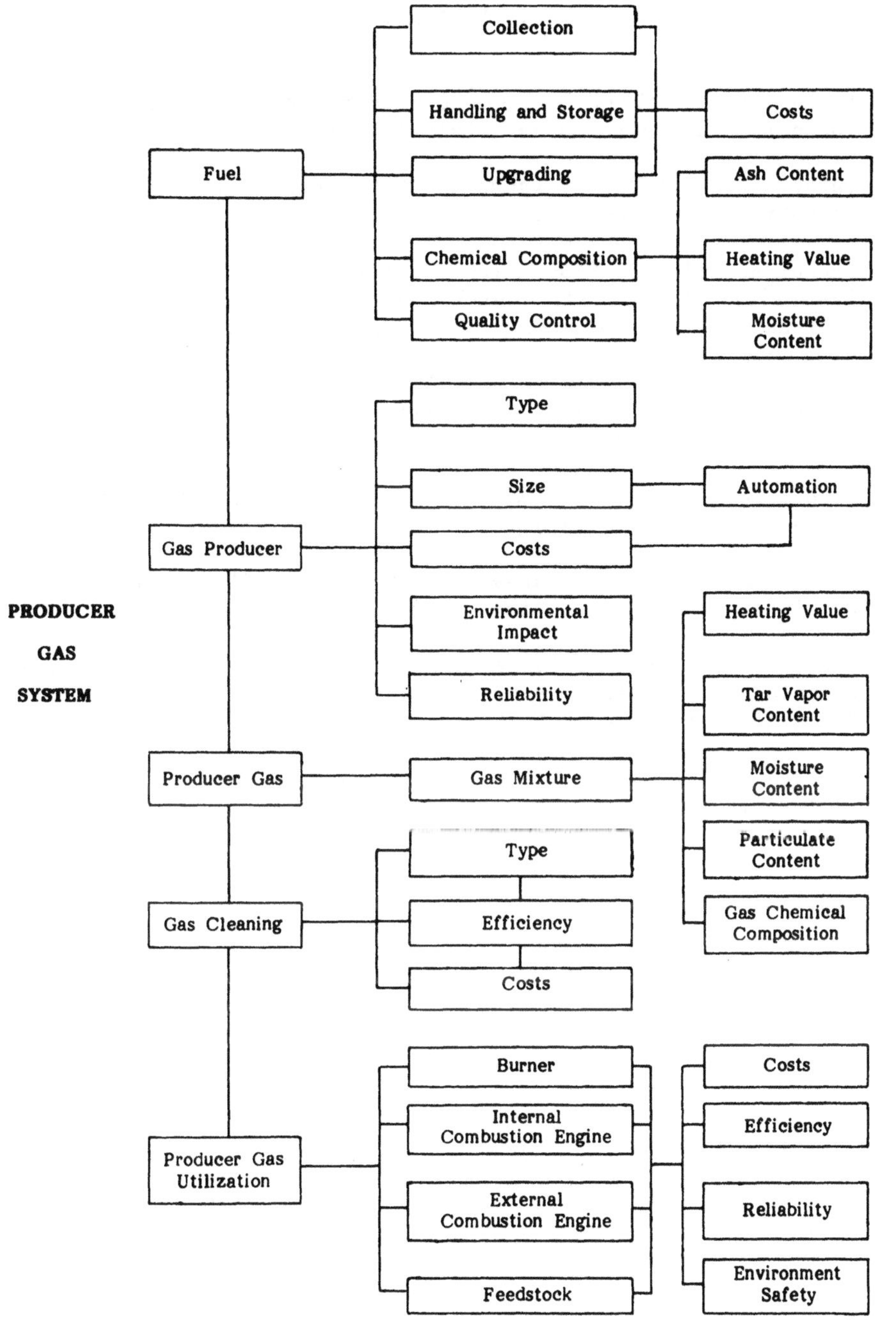

Figure 1-7. Block diagram of parameters involved in the gasification process.

I. REFERENCES

1. Bugge, G. 1927. Industrie der Holzdistillationsprodukte. Theodor Steinkopff Company, Leipzig, East Germany.

2. Harvey, R.G. 1982. Polycyclic Hydrocarbons and Cancer. American Scientist. July-August.

3. Kaupp, A. and J.R. Goss. 1982. State-of-the-Art Report for Small Scale (to 50 kW) Gas Producer Engine Systems. U.S. Agency for International Development, S&T/DIV, Room 509, SA-14, Washington, D.C. 20523.

4. Kennedy, W.B. 1940. Mixtures of Producer Gas and Petrol. Institution of Engineers 12(9):259-263. Australia.

5. Probstein, R. and R.E. Hicks. 1982. Synthetic Fuels. McGraw-Hill Book Company.

6. Reed, R.D. 1981. Furnace Operations. Third Edition. Gulf Publishing Company, Houston.

7. Reed, T.B. 1981. Survey of Biomass Gasification, Volume 2, Principles of Gasification. Publication #SERI/TR-33-239, Solar Energy Research Institute, Golden, Colorado.

8. Wyer, S. 1907. A Treatise on Producer Gas and Gas Producers. Hill Publishing Company, New York.

2. OBJECTIVES AND SCOPE

2.1 Objectives:

2.1.1 Determine the physical and chemical properties of rice hulls affecting the gasification process.

2.1.2 Develop a small rice hull gas producer, taking into consideration the unique physical and chemical properties of rice hulls.

2.1.3 Obtain performance data for the gas producer.

2.2 Scope:

The following physical and chemical properties of the rice hulls used in this research were determined:

2.2.1 True density of rice hulls.

2.2.2 Apparent density of rice hulls.

2.2.3 Bulk density of rice hulls.

2.2.4 True density of rice hull pellets.

2.2.5 Apparent density of rice hull pellets.

2.2.6 Bulk density of rice hull pellets.

2.2.7 Void space fraction of rice hulls.

2.2.8 Void space fraction of rice hull pellets.

2.2.9 Pore volume of rice hulls.

2.2.10 Pore volume of rice hull pellets.

2.2.11 Apparent surface area of rice hulls.

2.2.12 Apparent surface area of rice hull pellets.

2.2.13 Ash content of rice hulls and range for data from the published literature.

2.2.14 Ultimate analysis of rice hulls and range for data from the published literature.

2.2.15 Ultimate elemental analysis of rice hull ash and range for data from the published literature.

2.2.16 Softening and melting points of rice hull ash and reported phase diagrams.

2.2.17 Average weight of a single rice hull.

2.2.18 Microscale appearance of rice hulls before and after the gasification process.

2.2.19 Slagging and caking behavior of rice hulls and pellets.

2.2.20 Flame temperatures of the combustion and gasification process for rice hulls.

2.2.21 Tar cracking in a rice hull fuel bed.

2.2.22 Tar cracking in a pelleted rice hull fuel bed.

2.2.23 The pyrolysis products of rice hulls in a temperature range of 250-450 C:

Solid residue and composition

Tar generated

Water generated

Gas generated and composition

Carbon bound in the condensable phase.

2.2.24 Pressure drop across a rice hull and char bed as a function of the superficial gas velocity.

2.2.25 Size distribution of rice hulls.

2.2.26 Size distribution of the residues of rice hull gasification.

The following parameters characterizing the design and performance of a rice hull gas producer were determined:

2.2.27 Ash removal system: design and parameters involved in the design.

2.2.28 Design of the gas exit and parameters involved such as superficial gas velocity and solid particle entrainment.

2.2.29 Design of the combustion area of a rice hull gas producer to optimize temperature profiles and flow fields.

2.2.30 Fuel feeding systems for a rice hull gas producer.

2.2.31 Gas cleaning train for producer gas.

2.2.32 Gas composition as a function of gas output and furnace temperature.

2.2.33 Solid particle content of the raw gas stream.

2.2.34 Producer gas output versus size of the rice hull gas producer.

2.2.35 Adiabatic flame temperature of producer gas assuming various losses.

3. HISTORY OF GAS PRODUCER ENGINE SYSTEMS

Introduction

The present and future high energy prices which have affected Developing Countries most have made the possibility of gasification of waste products such as rice hulls and other crop residues a popular choice in Third World Countries and elsewhere. The alleged advantages of gasification seem to be intriguing for several reasons:

a) Decentralized energy conversion system which operates economically (at present high fuel prices) even for a very small scale of 5-10 hp output.

b) Gasification has been practiced for over 100 years and was a very well-established technology.

c) Producer gas was and still is generated in very simple stove-like combustors on a small scale.

However, the history of gasification provides a realistic picture of this technology. It is outlined in detail in the following section to identify the considerable social and economical problems that prevailed throughout its development and use. This chapter is taken from a publication in 1982 (Kaupp and Goss) which was based on the analysis of 63 original publications that dealt in length with the social, political and economical aspects of gasification prior to 1950.

History

The history of gasification can be dated back far earlier than usually stated. In 1669 Thomas Shirley conducted crude experiments with carbureted hydrogen and 30 years later Dean Clayton obtained coal gas from pyrolytic experiments. The first patents with regard to gasification were issued to Robert Gardner and John Barber in the years 1788 and 1791. Robert Gardner suggested the application of waste heat of furnaces to raise steam, by combusting the heated products in a boiler. However, the first confirmed use of producer gas from coal was reported in 1792. In this year Murdock generated gas from coal and used it to light a room in his house. For many years, after Murdock's development, coal gas was one of the principal fuels used for lighting purposes in England. It use declined in favor of electricity but the use of producer gas still continued and became increasingly important for cooking and heating. Experiments to gasify wood or at least use the gases obtained from charring of wood started surpisingly early (in the year 1798) when Lebon tried to gasify wood. In 1801 Lampadius proved the possibility of using the waste gases escaping from charring of wood. The process of generating water gas by reaction of water with a hot carbon bed was mentioned by Fourcroy in 1804. It took five more years before it was realized by Aubertot that the stack gases of blast furnaces can be combusted and used to roast ore and burn lime. He received a patent for this process in the year 1812. The first gas producer built used oil as a fuel and the patent was given to J. Taylor in 1815 who designed and operated the unit. Between the years 1815 and 1839 many patents were issued for utilization of waste heat and stack gas from blast furnaces. However, the first commercially used gas producer can be attributed to Bischof who built a large unit at the iron works of Audincourt, France in 1840. During the next 20 years many researchers and engineers improved the technology. They used low grade fuel and combusted the gases in gas fired furnaces. The real breakthrough came in 1861 with the Siemens gas producer which is considered to be the first successful commercial unit. Before the turn of the nineteenth century there are three more important events to mention. First, the introduction of the Dowson gas producer in 1878 which was the starting point of the modern gas producer-engine system. This was the first producer that was successfully used for stationary power engines. Second, the introduction of the Mond by-product process on a large scale in 1889. And third, the introduction of the Bernier suction gas producer in 1895, which was the beginning of the use of gas producers in small, compact units. The Mond by-product process proved for the first time that other valuable products such as ammonia could be obtained via gasification. The residual gas from this process was low in heating value but still could be used for industrial heating purposes. This process was also adapted to gasify highly volatile fuels such as peat and brown coal and several plants were in operation in Japan, the United States and Europe.

As far back as 1819 a portable gas producing apparatus comprising of a gas producer and a gas vacuum engine were patented in England. No record can be found that it was ever fitted on a vehicle. The task to actually operate a passenger vehicle with producer gas for the first time ever must therefore be credited to J. W. Parker who covered over 1000 miles with his 2 hp automotive gas producers in Scotland during the years between 1901 and 1905. It is interesting to note that the inadequate protection Bernier got for his patented gas producer-engine system, permitted other enterprising engineers the opportunity of directly benefiting from his system. Many competing designs were put on the market in increasing numbers for the next 15 years. One such make is the Brush Koela plant that was first introduced as a patented device in 1901 and was actually designed for export to India and other Developing Countries. The name Koela is the Indian word for charcoal. The oil engines used during this time period were actually replaced by producer gas engines. Some companies in England did a brisk business selling producer gas-engine sets to generate electricity throughout the country for domestic lighting. The necessity to stay ahead of competitors led some companies to utilization of the waste heat and the CO_2 generated in the process. However, these early attempts of co-generation were not very successful, although the general ideas behind it are no different from today's principles of co-generation. The first decade of the 20th century was also full of attempts to spread the new concept of suction gas producer-engine systems to other applications.

The Duke of Montrose convinced the British Admirality to introduce some of the new compact suction plants on ships, because similar experimental units were already in use on barges for channel and river transport in Germany and France. A small gas producer carried by four men and used for disinfection purposes was manufactured by J. Pintsch. The gas, rich in carbon monoxide, was used for killing mice, rats, or other vermin on farms and ships. The technology of gasification of wood and charcoal was stepped up, mostly to provide the colonies of the British and German Empires with gas producers that did not depend on scarce anthracite coal. H. A. Humphrey had considerable success with operating huge pumps on producer gas. Several types of these 1000 hp waterpumps were built in Alexandria (Egypt), Berlin (West Germany) and Chingford (England). Some enthusiasts considered producer gas the future fuel for internal combustion engines. On the other hand a talk given by Ade Clark for the Institution of Mechanical Engineers, London, in which he discussed industrial applications of the diesel engine signaled, in 1904, the increasing interest in this new diesel technology. The manufacture and operation of producer gas plants was in no way restricted to European countries and their colonies. In fact the United States Geological Survey had for several years investigated the economical value of coals and lignites as gas producer fuel. The early tests done with a pilot plant erected at the Louisiana Purchase Exposition in 1904 were very encouraging and demonstrated the use of many coals that could not be combusted in the existing steam-power plants. The fact that the technology of large updraft gas producers became more and more reliable encouraged gas engine manufacturers to build larger and larger units. Before the wide spread use of producer gas only small gas engines up to 75 hp were found economical to operate with town gas. However, the cheap producer gas led to the operation of huge gas engines. The first 600 hp engine was exhibited in Paris in 1900. Larger engines, up to 5400 hp were put into service in the U.S. shortly thereafter. The results of a survey of 70 plants out of the 376 existing plants in the U.S. in the year 1909 are published in United States Geological Survey, Bulletin #416.

With regard to the present situation, the USGS report is important because it states for the first time the many difficulties caused by lack of knowledgeable engineers, lack of knowledge and confidence in the technology on the part of the public, inexperienced salesmen not familiar with the details of the engine and the gas producer concept, lack of types of gas producers that could gasify inferior fuels and the large number of unsuccessful or only partly successful installations made during the experimental period of this development. One of the key problems with gas producer systems that has persistently remained to the present is quoted from the bulletin:

> "It cannot be denied that many of the difficulties charged to producer-gas power plants are due entirely to incompetent operators. Some plants have been put out of commission temporarily by the prejudices or the lack of ability and training of the operators or engineers in charge. A few of these failures are due to the impossiblity of finding men competent to operate the plants, but many of them have undoubtedly been the result of a short-sighted policy on the part of some manufacturers, who are not willing to give proper and necessary information about the design, construction, and operation of the plants made by them. The possibility of a sale at the time is apparently the only interest they keep in mind, and the future is allowed to take care of itself".

Sales brochures from many countries and personal contacts indicate the situation is very much the same today. The demand for better education of the designers and builders of gas producer plants and furnaces, drivers of automotive gas producer vehicles, the existence of special schools teaching gasification and the demand for higher wages for drivers of automotive gas producer vehicles can be found throughout the entire literature covering the 100 years of commercial gas producers.

Further development of the automotive gas producer was done by Porter and Smith in England during the First World War. The impetus for this work was the possiblity of disruption of gasoline supplies which had become the dominant fuel for motor transport. Although most of the early development of automotive gas producers was done in England, wide spread application during and after the First World War was crippled by the British taxation system that assigned taxes to cars according to their weight which included the gas producer. The 1919 special report of the Inter-Departmental Committee on the employment of gas as a source of power which dealt at considerable length with the automotive gas producers and its advantages was not followed by any government action to put the automotive gas producer in a more favorable tax situation.

A totally different situation prevailed in France. There the use of wood and charcoal as a fuel had a long history and the French government was actively encouraging the development of automotive gas producers after 1919. Further public awareness of this method to drive an automobile was greatly increased through ralleys organized each year since 1926 by the Automobile Club de France. The distances that had to be covered were between 1600 and 3000 km. One of the greatest names in the development and manufacture of automotive gas producers was the Frenchman, Imbert. He filed his first patent for a downdraft gas producer in 1923 and many successful designs including the recently built small automotive gas producers are based on this design. The interest in the automotive gas producer faded in France during the 1930s and most of the development in this field continued in Germany. In fact the Imbert Company is again manufacturing gas producer-engine systems in West Germany. Although the automotive gas producer never played any role in the development of gasification in the U.S., more than 12,000 stationary gas producers were in operation during the 1920's and 1930's in the U.S. and Canada. Over 150 companies in Europe manufactured small and large gas producers for various applications. The gas producer concept was especially appealing for applications in remote areas or Developing Countries which had brush or timber. For instance, the British company, Crossley, sold gas producers for remote mines in Australia and the Tulloch Reading 50 hp truck developed in England was mostly purchased by the Empire Cotton Growing Cooperation for use in Nigeria.

The next decade from 1930 to 1940 can clearly be considered as the development period for small automotive and portable gas producers that reached its peak during World War II. New concepts and designs such as crossdraft gas producers were developed or improved. Efforts were undertaken to build the automotive gas producers lighter and improve the gas cleaning system which was the vulnerable part of the units. New units, capable of gasifying more readily available fuels such as bituminous coal, anthracite and wood, were developed and tested in small numbers. The British gasification efforts were still more directed to their overseas markets and not so much for domestic use. There were signs of an increasing critical view toward the automotive gas producer in France. It was claimed that at least one new gas producer mounted on a truck was more expensive to run and operate than a comparable gasoline truck despite all government grants and subsidies. It is of interest to recall the official position of the French and British governments during the early 1930s. Authorities in both countries felt at that time that the automotive charcoal gas producer was more suitable for their colonies where the supply of gasoline was scarce, and wood that could be made into charcoal at very low labor costs was readily available. The emerging gas producers using wood and low grade coal were not given much of a chance for general use. History has proven that assessment to be correct.

The first well reported conversion of internal combustion engines, in this case tractors, to producer gas drive under economic pressure occurred during the 1931 to 1934 period in Western Australia. The large quantities of wood available, the negligible oil resources at this time and the collapse of the wheat prices during 1930's set the scenario for a rather hasty, uncoordinated conversion of kerosene tractors to producer gas operation. Many farmers, in order to avoid bankruptcy had to consider all alternatives, including producer gas, although it was well known that the power loss of the tractors would be considerable. What happened during these years until the recovery of the wheat prices was just a small part of what happened later during World War II on a much broader basis. Many gas producers were failures from the start. Others deteriorated rapidly owing to faulty construction. Several firms were interested in the manufacture and sale of such units, but had neither the money nor time to do the necessary research and development engineering. As a consequence, there were often totally dissatisfied customers, who after a short trial, resolved they would never again have anything to do with gas producers.

On the other hand, a small number of farmers having ingenuity and mechanical skill, operated their units very satisfactorily for a number of years. In this context it should be mentioned that there has never been an automotive engine especially designed and built for producer gas, although the technology was wide spread for over 100 years. With plentiful fossil fuels available during peaceful and stable economical times, there was no need for the producer gas concept. During emergencies and war times the concept of producer gas engine systems was always so hastily brought back that there was simply not enough time and money available to develop a specially designed producer gas, internal combustion engine for automotive use. This explains in part the difficulties some farmers had in converting their kerosene tractors to producer gas drive. The interest in gas producers faded quickly after the 1930's depression was over. Only 62 producer gas tractors out of 4548 tractors in Western Australia were operating at the end of 1937.

In late 1930's the effort of Nazi Germany to accelerate the conversion of vehicles to producer gas operation was the beginning of a worldwide effort to use the gas producer concept as part of a plan for national security, independence from imported oil and acceleration of the agricultural mechanization. A typical example was the Soviet Union. The build-up of the military as well as rapid expansion of heavy industry necessitated a major change in the mechanized agricultural units. The change was directed toward the fuel used. It became apparent that despite a high priority for the agricultural sector, the transport of the fuel was becoming a problem. The big agricultural areas were far from the large oil fields and the distribution of the fuel even when plentiful was one of the biggest problems. The introduction of gas producer powered tractors and trucks to the Russian farmers can therefore not be viewed as an emergency measure to reduce the consumption of gasoline and diesel oil. Instead it was viewed as an alternative to using fuels available locally and thereby easing the transportation and distribution problem. Almost all early Russian tractors were powered by gasoline engines which required extensive rebuilding of the engine to avoid a severe power reduction. (A later model, the Stalinez C65 tractor, and the Kharkov caterpillar tractor were equipped with diesel engines). From the design of the gas producer and its gas cleaning system, it seems most likely that various German gas producers were used as the basic design for these models. Despite some criticism about the gas producer concept, its economics and future, new advanced crossdraft gas producers were built in France. In particular the Sabatier and Gohin Poulence plants showed an astonishing performance, equal to most gasoline powered vehicles. However, it became more and more obvious that good gas producer performance was closely connected to the quality of the fuel. Plants like the Sabatier, or later the Swedish Kalle model, were highly reliable and worked well only with specially manufactured charcoal having carefully controlled quality. In 1938 most European countries stimulated the use of producer gas through subsidies for conversion, favorable tax benefits or even edicts such as in France that required all public transport companies to change at least 10% of their vehicles to producer gas. The Italian government was even more strict, requiring all buses in public service to use home produced fuel, wood charcoal, alcohol or home produced petrol and oil. These various measures led to 4500 gas producer vehicles in France, 2200 in Germany and over 2000 in Italy by the early part of 1939. England, the country that did most of the pioneer work in the beginning, however, saw its producer gas program entangled in politics, resulting in very little conversion to producer gas for vehicles. We learn this from an article written by the Coal Utilization Council appearing in the Fuel Economist in July 1938. The Director of this organization complained bitterly about the stubborness of the British government in this matter. Nevertheless, some British bus companies ran their city buses on producer gas quite successfully and on schedule.

What happened to the development of the automotive gas producer after 1939 must be seen in the context of the World War II. From the number of articles published about gasification in German journals each year and the work of several national committees on the subject it was obvious that Germany was much better prepared to deal with the logistic problems associated with the operation of hundreds of thousands of automotive gas producers. However, the most drastic development took place in Sweden, which experienced a most severe fuel shortage. Other countries delayed the conversion to producer gas operation, because there was simply no need for it. For instance, not many automotive gas producers were seen in Australia in the year 1940, compared to a considerably larger number in New Zealand which was much earlier affected by the fuel shortage. The United States coped with gasoline shortage by means of rationing but nevertheless automotive and stationary gas producers were manufactured in Michigan. They were not available for domestic use and most of them were sold to China under Lend-Lease terms.

"Woman Who Fled Nazis Makes Gas Producers in Michigan Plant for Export to China" was one of the headlines of several articles that appeared in the National Petroleum News and Chicago Tribune about this activity.

The development of the European gasification activities was closely monitored by the Forest Service of the United States Department of Agriculture and some of the findings have been published. At the end of 1944 it was concluded that wide spread commercial adoption of gas producers in the United States would not be promoted. Only under special circumstances in remote areas, gas producer operation might be acceptable.

Even after the outbreak of the war, the British government was in no hurry to regulate or require the use of automotive gas producers. One of the reasons was the unsuitability of most existing gas producers for the soft and brown coals of England, which had little anthracite. Nevertheless, a so-called government emergency crossdraft gas producer was developed especially for the British coals and low temperature coke and it was planned to manufacture 10,000 units. The government developed producer worked reasonably well but in 1942 it became increasingly difficult to obtain the necessary low ash coal to run the gas producer and plans to mass produce the unit were given up. The conversion of vehicles to producer gas drive was therefore mostly restricted to bus companies and some private companies that installed the stationary Cowan Mark 2C gas producer as an emergency power supply to factories affected by air bombing. Therefore, large-scale conversion of vehicles took place in Sweden and the countries occupied by Germany during World War II.

In December, 1939, about 250,000 vehicles were registered in Sweden. At the beginning of 1942 the total number of road vehicles still in service was 80,000, about 90% of which were converted to producer gas drive within 1 1/2 years. In addition, almost all of the 20,000 tractors were also operated on producer gas. Forty percent of the fuel used was wood and most of the remainder charcoal. Dried peat was used to some extent. This fast and almost complete conversion was accompanied by the drastic decline of imported petroleum from 11 million barrels in 1939 to 800,000 in 1942.

It is far more interesting to recall the logistic difficulties associated with the conversion of gasoline vehicles on a large scale during World War II, because the technical advances made after 1940 were not significant and dealt mostly with the improvement of gas cleaning systems and better alloys for the gas producer shell.

Schlapfer and Tobler, who conducted extensive tests with various gas producers during the 1930 to 1939 period in Switzerland, pointed out the human element involved. They argued that most of the converted post buses running on producer gas in Switzerland did not perform well because drivers had difficulties getting used to the new driving style and were reluctant to do the additional work involved. Most troublesome was the required daily cleaning of the entire cleaning system and the preparations for the next run, which included the clearing of the fuel hopper, because overnight storage of the fuel in the fuel hopper caused considerable starting difficulties due to moisture buildup. It also became apparent that neither the manufacturers nor the general public really understood the problems associated with a gas-producer operated bus.

One company in France had a mixed fleet of gasoline and gas producer driven vehicles. The drivers had to carry out refueling and making repairs after the day's work at regular rates of pay. The producer vehicles were constantly having to go to the shop for repairs which the drivers alledged were beyond what they could do. After the company decided to pay overtime for the time spent to clean and refuel the gas-producers, the producer vehicles became at once as trouble free as the gasoline vehicles.

This situation also existed in Germany prior to World War II. In the early 20th century, owners of large stationary gas producers talked about sending their engineers to special schools on gasification and paid them higher wages to ensure that the gas producer was properly operated. The German government finally agreed after many years of rejection, to pay their drivers of automotive gas producers higher wages, which improved the situation. However, the uninformed private driver remained a persistent problem. At the beginning he was faced with hundreds of makes of gas producers and no manufacturer's guarantee about the performance. Although one could not prove that some manufacturers actually sold equipment they knew would not work, it cannot be denied that many of them did not know much about the performance of their units or could only prove reliable performance with high quality fuel having carefully controlled physical and chemical properties. Large numbers of unsatisfied customers finally led to government action in Germany and Sweden as well as in the occupied countries. The number of manufacturers of gas producers was significantly reduced to about 10 with models that had been proven to be successful. However, the fuel supply and the quality of gas producer fuel was still a problem that actually was never solved. Until the end of 1941, wood and charcoal were the fuels most widely used in Germany. The collection and preparation of gas producer fuel was handled by the Gesellschaft fur Tankholzgewinnung und Holzabfallverwertung which kept a tight control on the size, shape and moisture content of the fuel. The fuel could be purchased at over 1000 official filling stations all over the country. This service was more or less operated and organized like today's oil companies and gasoline stations.

It soon became apparent that at the prevailing wood consumption rate and the tendency of drivers to use charcoal, there would not be much forest left within a few years. The construction of charcoal gas-producers was therefore forbidden in France and Denmark after July 1, 1941 and greatly restricted in Germany and Sweden. The new policy was to encourage the use of brown-coal, peat coke, anthracite and low temperature coke. It however can be concluded that their use was plagued by problems with the quality of the fuel, such as high sulfur content, too much volatile matter, poor physical shape of the various cokes sold, too expensive production methods and improper handling of the fuel bags. Most customers did not understand the differences among the various fuels they could buy or their influence on the gas producers. The situation today is about the same and any introduction of small stationary or portable gas producers on a broad basis would likely lead to the same difficulties. Some users of automotive gas producers even produced their own fuel out of brush wood collected in the national forests.

A slightly different situation prevailed in Sweden with its vast supply of wood. At the beginning the unrestricted use of charcoal led to various designs of high performance gas producers, which operated very well as long as they were fired with the specially prepared charcoal they were designed for. The tar oils from wood carbonization were also not wasted and used for heavy fuels and as a lubricant. Over 3000 furnaces producing charcoal were in operation in 1944, to provide the necessary fuel for metallurgical operations and the fleet of gas producers. Although the officially produced fuel was strictly classified and quality controlled, not all of the fuel related problems could be solved. For instance, first grade low volatility fuel of less than 3% volatiles turned out to be medium volatility fuel with over 8% volatiles that could not be gasified in most gas producers. Hard, high grade charcoal leaving the factories with a low moisture content of 10% and only a 10% fraction of fines, reached the consumer broken up and crumbled with a moisture content of over 20% and was therefore rendered useless. Although the emergency situation was on everybody's mind, the temptation was high to buy and operate the very convenient, high performance gas producers which depended on special fuels.

It is obvious that an automotive gas producer that can be started within 2 minutes and does not require much cleaning sounded much more appealing for the private customer than one with more flexibility with regard to the fuel needed to operate the unit. The tendency to modify the fuel for a gasifier in question instead of investing the time and money to design and construct a gas producer for a fuel in question can be found throughout the entire history of gasification. This approach was not changed during the first 100 years of gasification and present signs indicate that there will be slow progress toward designing gas producers for specific fuels.

Although the number of accidents related to the use of automotive gas producers was considerably higher than with gasoline vehicles, most accidents were due to negligence of the driver. The increasing number of accidents caused by operators not familiar with their equipment was of much concern to the Swedish government and the manufacturers. This was reflected in very detailed operation manuals and the introduction of a special driver's license for the operation of an automotive gas producer. Of concern were simple operational mistakes such as not ventilating the unit after a day's use which resulted in a gas build up in the gas producer that could explode while the owner was checking the fuel level next morning. Other operators had the opinion that as long as the engine was running on the produced gas everything was fine and switched too early to producer gas drive during the startup period. In most cases this led to totally tarred up manifold and valves, because the initially produced gas, although of high heating value was rich in higher hydrocarbons that condensed in the engine. More serious and not so easily controlled is the danger of accumulative carbon monoxide poisoning which occurred frequently according to Swedish reports. The problems in the past with automotive gas producers, should be viewed in the light of the enormous task that was undertaken in Europe to convert hundred of thousands of gasoline vehicles to producer gas drive within three years in a difficult time. An automotive gas producer must also be viewed as the most advanced gas producer, much more difficult to design and operate than a stationary unit.

Shortly after World War II, automotive gas producers as well as all the large stationary units were put out of service because of abundant, cheap supplies of gasoline, diesel oil, natural gas and electricity. The change away from producer gas operation was also drastically reflected in the research done in this field. The number of publications listed in major engineering indexes dropped sharply from several hundred a year to less than 10 a year during the 1950 to 1970 period. It can be said with few exceptions, gasification and in particular small portable gas producers were a forgotton technology during this time period. Research done in this field, which can be called a considerable contribution to the advancement of automotive gas producers, took place in Sweden during the 1957 to 1963 period. This research was initiated by the Swedish Defense Department after the Suez Crisis and undertaken by the National Machinery Testing Institute. The research made considerable contributions to the improvement of the gas cleaning system and the modifications of diesel engines for gas producer drive.

The 1970s brought an increasing renewed interest in this form of power generation and a more general look at the complexity of gasification. Some of the present work concentrates on the revival of the old ideas and designs and their modification and expansion to fuels different from wood and coal. A worldwide search for small scale gas producers in operation and researchers working on the subject as well as the increasing number of daily inquiries about gasification received, show a considerable interest and demand in small gas producers. However, the idea still persists that gas producers are a simple stove-like energy conversion system that is easy to design and operate. The present demand is therefore also stimulated by the belief that gasifiers can convert almost any carbonaceous material to useful mechanical and electrical energy. This image of a gasification system is far removed from any reality and in particular the history of gasification has shown that a fixed bed gasifier providing fuel for an internal combustion engine is a very selective energy conversion system with little flexibility with regard to the fuel it was designed for. A further handicap is the little knowledge we have about the behavior of various biomass fuels under thermal decomposition. This knowledge is certainly basic for any further optimization of gas producers and cannot be obtained easily. On the other hand, amazing performances of gas producer-engine systems have been reported and verified throughout the history of gasification. It is not just an assumption but confirmed reality that trucks have been operating on producer gas for over 300,000 km with no major repair and less engine wear than that resulting from using only diesel fuel. Large Italian rice mills have gasified their rice husks and used the gas to drive the power units used for milling for decades. The number of quite satisfied owners of small and large gasifiers is certainly not small and there is lots of evidence that it can be done. The history of gasification has also shown that it is not one of the most convenient technologies.

3. REFERENCES

1. Allcut, E.A. 1942. Producer Gas for Motor Transport, Engineering Journal 5(4):223-230.

2. Anderson, M.. 1938. Case for the Encouragement of the Producer-Gas Vehicle in Britain. Fuel Economist 14:245-246, 256-257.

3. Anonymous. 1933. A New Gas Producer-Gas Plant for Road Transport. The Commerical Motor, 20:787-788. January.

4. Anonymous. 1939. Alternative Fuels for Wartime. Gas and Oil Power. pp. 235-238. October.

5. Anonymous. 1941. Emergency Gas Supplies for Factories. Power and Works Engineer 36:137-139. June.

6. Anonymous. 1938. Forest Gas for Traction. Engineer 166(4311): 230-231.

7. Anonymous. 1939. Gas as a Substitute for Gasoline Part 1. Petroleum Times 42(1073:169-170, 189.

8. Foster Wheeler Energy Corporation. 1978. Gas from Coal: A Volatile Solution. Energy Guidebook. pp. 108-110.

9. Anonymous. 1939.· Gas Producer for Road Vehicles. Engineering. pp. 631-632. May 26.

10. Anonymous. 1939. Gas Utilization for Automobiles. Gas Age. December 7.

11. Anonymous. 1943. German Portable Gas Producer Practice. Engineering 155:423-424. May.

12. Anonymous. 1949. Improvements in the "Brush Koela" Gas Producer. Engineering 169(4398):395.

13. Anonymous. 1940. Specification for Portable Gas-Producer Fuels. Engineering. p. 150. February.

14. Anonymous. 1942. Producer Gas Plant Manufacture. Gas and Oil Power 37(443):147-150.

15. Anonymous. 19 . Producer Gas: Present and Future. Gas and Oil Power 40(473):49-50.

16. Anonymous. 1943. Producer Gas versus Petrol Operation in Germany. Petroleum Times 47(1193):190.

17. Anonymous. 1942. The P.S.V. Gas Producer. Bus and Coach. pp. 228-230. November.

18. Anonymous. 1945. The Soviet Producer-Gas Tractors. Gas and Oil Power. pp. 89-95. March.

19. Anonymous. 1929. The Tulloch-Reading Gas Producer for Motor Vehicles. Engineering 127:641-644.

20. Bailey. M.L. 1979. Gas Producers for Motor Vehicles: A Historical Review. Department of Scientific and Industrial Research, Chemistry Division, Report CD 2279. New Zealand.

21. Branders. H.A. 1941. Producer Gas is the Motor Fuel of Finland. Automotive Industries. pp. 482-485, 522-523. May.

22. Breag, G.R. and A.E. Chittenden. 1979. Producer Gas: Its Potential and Application in Developing Countries. Tropical Products Institute, Report G130, London, England.

23. Brownlie, D. 1940. Producer-Gas Driven Vehicles. The Iron and Coal Trades Review. pp. 121-123.

24. Campbell, J.L. 1942. Gas Producers: An Outline of the Compulsory Government Tests in Australia. Automobile Engineer 32(422):156-158.

25. Clarke, J.S. 1940. The Use of Gas as a Fuel for Motor Vehicles. Institute of Fuel Journal 13(70):102-117.

26. Dimitryev, A.P. 1940. Automotive Gas Generators Used in USSR. Automotive Industries 83(10):534-535, 551.

27. Dunstan, W.N. 1945. Gas Engine and Gas Producer Practice in Australia. Engineer 180(4688):400-401.

28. Egloff, G. and M. Alexander. 1944. Combustible Gases as Substitute Motor Fuels. Petroleum Refiner 23(6):123-128.

29. Egloff, G. 1947. Fuels Used in Sweden. Petroleum Engineer 18(5):86-88.

30. Forbes, W. 1939. Experiments with Gas Producer Vehicles in Cardiff. Passenger Transport Journal. pp. 201-205. November.

31. Fowke, W.H. 1938. Operating Results with Producer Gas. Bus and Coach 10(2):84-86.

32. Freeth, E.E. 1939. Producer Gas for Agricultural Purposes. Journal of the Department of Agriculture of Western Australia 16(4):371-414.

33. Gall, R.L. and J.D. Spencer. 1966. Caking Coal Behavior in Gas-Producer Tests. Coal Age 71(2):128-130.

34. Goldman, B. 1938. Fuels Alternative to Oil for Road Transport Vehicles. Fuel Economist 14(248-252. July.

35. Goldman, B. and N.C. Jones. 1939. The Modern Portable Gas Producer. Institute of Fuel 12(63):103-140.

36. Goldman, B. and N.C. Jones. 1938. The Modern Portable Gas Producer. The Engineer 166:248-252. December.

37. Goldman, B. and N.C. Jones. 1939. The Modern Portable Gas Producer. The Petroleum World 36(460):3-5.

38. Greaves-Walker, A.F. 1916. The Design and Construction of a Producer-Gas House for Clay Plants. Transactions of American Ceramic Society 18:862-866.

39. Hurley, T.F. and A. Fitton. 1949. Producer Gas for Road Transport. Proceedings of the Institution of Mechanical Engineers 161:81-97.

40. Kralik, F. 1943. Rail Car with Charcoal Gas Producer. The Engineers' Digest. pp. 24-25. December.

41. Lang, W.A. 1943. Alternative Fuels for Motor Vehicles. Engineering Journal 26(8):449-454.

42. Langley, F.D. 1942. The Revival of Suction-Gas Producer. Gas and Oil Power 37(446):236-240.

43. Lindmark, G. 1944. Swedish Gas Producer Buses. Bus and Coach. pp. 266-269. April.

44. Littlewood, K. 1977. Gasification: Theory and Application, Progress in Energy and Combustion Science 3(1):35-71.

45. Lustig, L. 1947. New Gas Producer for Dual Fuel Engines. Diesel Progress 13(5):42-43.

46. Mellgren, S. and E. Andersson. 1943. Driving with Producer Gas. National Research Council of Canada RP 15/43. Ottawa, Canada.

47. Miller, R.H.P. 1944. Gasogens. U.S. Department of Agriculture, Forest Service, Forest Products Laboratory. Madison, Wisconsin.

48. Overend, R. 1977. Wood Gasification: An Old Technology with a Future? Sixth Annual Meeting, Biomass Energy Institute Symposium. Winnipeg, Manitoba, Canada. October 12.

49. Pavia, R.E. 1942. Woodgas Producers for Motor Vehicles. Institution of Engineers Journal 14(12):279-292.

50. Rambush, N.E. 1923. **Modern Gas Producers.** Van Nostrand Company. New York.

51. Renton, C. 1940. Producer Gas Tests in the Queensland Railway Department. Institution of Engineers Journal. pp. 274-278. Australia. October.

52. Ridley, C. 1944. Temporary Fuels, A Consideration of the Prospect of Their Permanency. Automobile Engineer 34(446):63-67.

53. Roberts, R.P. 1933. Producer Gas Equipment on Tractors in Western Australia. Journal of the Department of Agriculture of Western Australia 15(4):391-402.

54. Ruedy, R. 1944. Wood and Charcoal as Fuel for Vehicles. National Research Council of Canada (1157).

55. Skov, N.A. and M.L. Papworth. 1975. **The Pegasus Unit.** Pegasus Publishers Inc. Olympia, Washington.

56. Telford, W.M. 1949. Some Notes on the Design of Mobile Producer Gas Units. Institute of Engineers Journal 12(11):299-304. Australia.

57. Telford, W.M. 1941. Some Notes on the Design of Mobile Producer Gas Units. Gas and Oil Power 36:179-181.

58. Tookey, W.A. 1952.. Suction Gas Plant Development Fifty Years Ago. Engineer 193(5028):754.

59. Twelvetrees, R. 1944. Paving the Way for Producer Gas Operation. Bus and Coach. pp. 104-107. February.

60. Walton, J. 1940. Alternative Fuels. Automobile Engineering 30:91-92. March.

61. Woods, M.W. 1940. An Investigation of the High-Speed Producer Gas Engine. Engineer 169(4401):448-450.

62. Woods, M.W. 1938. Producer Gas Vehicles. Institution of Engineers Journal 10(3). Australia.

63. Wyer, S.S. 1906. **A Treatise on Producer Gas and Gas Producers.** Hill Publishing Company.

4. LITERATURE REVIEW

Gasification research and commercial production came to an almost complete standstill after World War II. The current revitalization of the technology is due to the growing belief that fossil fuel supplies will be exhausted within 50-100 years and that transitional energy technologies will be necessary until direct utilization of the sun's energy can be instituted on a large scale.

The present work done in the field of biomass gasification is insignificant when compared to the work done between 1840 and 1950. An extensive literature review was therefore conducted which identifed about 1200 publications of worthwhile content. More than 640 of these publications were obtained, analyzed and organized by the following keywords:

a) Internal Combustion Engine

b) Developing Countries

c) Gas Cleaning

d) Chemistry of Gasification

e) Portable Gas Producer Engine Systems

f) Stationary Gas Producer Engine Systems

g) Biomass Fuels

h) Fossil Fuels

i) Theory of Gasification

j) Economics of Gasification

The results have been published by Kaupp and Goss (1982). Also incorporated in the report is a survey conducted in 63 countries concerning present research efforts in this field. Of over 500 persons contacted,

148 in 34 countries either had experience with biomass gasification or are presently working on it.

Rice hulls are one of the most abundant crop residues in the world (an estimated 70 million tons in 1982). This byproduct of rice milling often causes disposal problems. Many studies have been undertaken with regard to the utilization of rice hulls for a wide range of applications. An excellent survey of the utilization of rice hulls as an energy source has been done by Beagle (1978). Books on the various applications have been published by The American Cereal Chemist Association (Houston, Chapter 12, 1972) and Angladette (1966). Ultimate analyses of rice hulls as well as ash content measurements are given by Beagle (1978), Houston (1972), Jones (1953) and Borasio (1928) for various locations. A number of 40 - 80 hp rice hull gas producers were in operation for decades in Italy. Descriptions and sketchy performance data are given by Garbarino (1911) and Francesco et al (1917) about the 80-100 hp Italian rice hull gas producers. The journal IL Giornale di Risicoltura has regularly published information on rice hull gasification as well as on the properties and utilization of rice hulls, most of which appeared between 1908 and 1950.

The German engine maker, Humboldt-Deutz Motoren AG, did extensive testing on rice hull gas producers during the 1930s. Possenti (1944) describes a successful Italian rice hull gas producer, named the Balestra unit after the engineer who built and operated it for over 30 years in the gas plant of Vercelli.

Tucker (1944) investigated some physical properties of rice hulls such as thermal conductivity and true density. However, his investigative methods leave some doubt about the validity of the published data.

The most recent reports of existing rice hull gas producers are listed in the FAO publication RAPA No. 61 (FAO Bangkok, June, 1982). The publication discusses rice hull gasification efforts in Malaysia, Thailand, The Peoples Republic of China and the Philippines.

The Egyptian government shut down their last remaining rice hull gas producer because one worker died of carbon monoxide poisoning while two others developed respiratory diseases (Beagle, 1982). A well-preserved (1925) rice hull gas producer and engine (Crossley-Premier) was reported by Beagle (1983) in Burma. It may therefore be concluded that rice hull gasification on a larger scale (40 - 150 hp) is well documented and was successful at least in Egypt, Italy and Burma. Recent work and reports about development units as well as laboratory units at the university level must be viewed with great caution. Well-documented tests on rice hull gasification in a fluidized bed are given by Van den Aarsen (1982).

4. REFERENCES

1. Angladetta, A. 1966. LeRiz. G.P. Maisonneuve and Larose, Paris.

2. Beagle, E. 1978. Rice husk conversion to energy. FAO Agricultural Services Bulletin #31.

3. Beagle, E. 1982. Personal communication. October, 1982.

4. Beagle, E. 1983. Personal communication. January, 1983.

5. Borasio, L. 1928. Sulla composizione chimica della lolla di riso. IL Giornale di Risicoltura, Anno XVII, No. 3.

6. Francesco, et al. 1917. Processo industriale per la utilizzazione della lolla. Premiata Tipo litografia Gallardi and Ugo, Vercelli 1917.

7. Garbarino, G. 1911. Utililzzazione integrale della lolla di riso come combustibile. November 1911.

8. Jones, J.D. 1953. New refractory from vegetable source. Canadian Metals. January.

9. Humboldt-Deutzmotoren AG. 1933. Gas producer for using vegetables wastes, e.g., rice husks, nut shells of cacao, coffee or palm kernels. Chem. Abstracts Vol. 27:5520.

10. Humboldt-Deutzmotoren AG. 1934. Gas producer for using husks and similar vegetable wastes. Chem. Abstracts, Vol. 28:3565.

11. Kaupp, A. and J.R. Goss. 1982. State-of-the-Art for small scale (to 50 kW) gas producer engine systems. U.S. Agency for International Development, Washington, D.C. 20523.

12. Possenti, A. 1944. Gasification plant for rice hulls in the gas plant of Vercelli. Risicolture Vol. 30:243-251 and Chem. Abstracts Vol. 37:3247.

13. Tucker, P.W. 1944. A study of the physical and chemical properties of rice hulls. M.S. Thesis, Louisiana State University, Department of Chemistry.

14. Van den Aarsen, F.B. et al. 1982. Performance of a rice husk fueled fluidized bed pilot plant gasifier. First International Producer Gas Conference, Colombo, Sri Lanka. November.

5. CHEMISTRY OF GASIFICATION OF RICE HULLS

Introduction

Gasification of coal and biomass had its heyday at a time when very little was known about the process except that incomplete combustion generated a gas which in turn could be further oxidized (burned).

The simple definition of gasification given earlier as a process that stores a maximum of chemical energy in the gas phase does not include any suggestions about how this is achieved, nor does it disclose the principles behind this conversion process.

Any air-blown combustion process yields a range of gases such as:

CO, CO_2, CH_4, C_2H_2, C_2H_4, C_2H_6, C_3H_m, C_4H_m, N_2,
H_2, $H_2O(g)$, NO_2, NO, O, O_2, N and H

as well as condensable vapors consisting of higher hydrocarbons.

Advanced computer programs as well as experimental results have shown quite clearly that at the prevailing gas producer temperatures of 800-1300°C the products which contribute more than 1% to the total gas volume are restricted to CO, CO_2, CH_4, N_2, H_2, $H_2O(g)$ and in some instances O_2.

One elemental chemical ultimate analysis of dry rice hulls shows that 100 kg of them can be represented by:

$$C_{3.45}\,H_{4.94}\,O_{2.33}\,N_{0.04}\,(SiO_2)_{0.26}$$ (see Table 6-1)

This representation accounts for 99% of the constituents of rice hulls. It is therefore up to the individual to decide how the gasification process will be characterized based on the given input and output species. One should realize that there are many options available and that those models which most closely resemble a real situation are in most cases not solvable. The non-solvability of models is not primarily caused by insufficient numerical methods. The major handicap and missing link are the physical parameters of the fuel and the design parameters of the gas producer that must be incorporated into the governing equations, either as dependent variables or as boundary conditions.

The complexity is shown by a list of properties involved in modelling the gasifier performance:

a) Temperature Field
b) Flow Field
c) Species Concentration Profiles
d) Intrinsic Reaction Rates of the Homogeneous Reactions
e) Boundary Conditions
f) Mass Transport across Phase Boundaries and Heterogeneous Surface Reactions.

Formation Reactions

It is customary to list the formation reactions of the major products as well as their interaction. One should not conclude that formation of these species can be adequately characterized by this simple set of equations (at 25°C, 1 atm):

Combustion	$C + O_2 = CO_2 - 394$ MJ/kg-mole	(5-1)
Water Gas	$C + H_2O = CO + H_2 + 131$ MJ/kg-mole	(5-2)
Water Shift Reaction	$CO + H_2O = CO_2 + H_2 - 41$ MJ/kg-mole	(5-3)
Boudouard Reaction	$C + CO_2 = 2\ CO + 173$ MJ/kg-mole	(5-4)
Methane Reaction	$C + 2H_2 = CH_4 - 75$ MJ/kg-mole	(5-5)

It is difficult to determine whether CO_2 is formed directly according to Equation 5-4 or whether it is the final result of a sequence of reactions such as:

$$C + 1/2\ O_2 \rightleftarrows CO \qquad (5\text{-}6)$$

$$CO + 1/2\ O_2 \rightleftarrows CO_2 \qquad (5\text{-}7)$$

In addition, methane can be formed in many ways in homogeneous reactions such as:

$$CO + 3\ H_2 \rightleftarrows CH_4 + H_2O \qquad (5\text{-}8)$$

$$CO_2 + 4\ H_2 \rightleftarrows CH_4 + 2\ H_2O \qquad (5\text{-}9)$$

The basic set of equations listed above does allow one to identify various zones in a gas producer as shown in Figure 5-1 for a downdraft gas producer where fuel and air are flowing in the same direction.

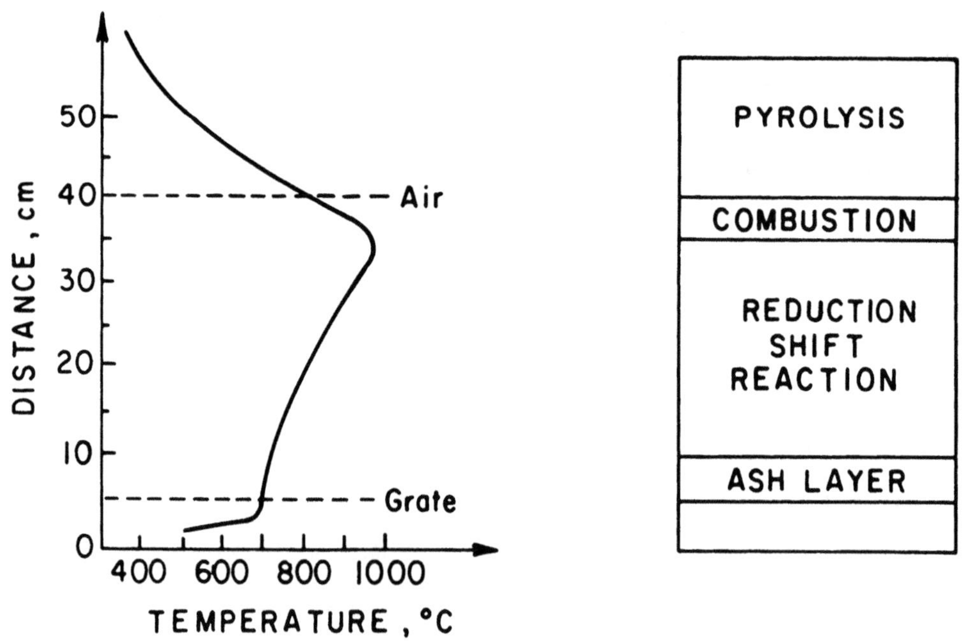

Figure 5-1. Co-current or downdraft gasification.

Reaction Zones

Splitting up a gas producer into strictly separate zones is certainly not realistic but is nevertheless conceptually useful. Most commonly we have:

a) Pyrolysis Zone

b) Combustion Zone

c) Reduction and Shift Zone

It is necessary to distinguish between reversible reactions which may be in equilibrium and irreversible reactions which can hardly be modelled. It is therefore difficult to find an adequate mathematical description of the processes taking place in the pyrolysis zone. The pyrolysis process, or low temperature decomposition of biomass, is highly irreversible. It involves the release of oxygen, hydrogen and nitrogen in the fuel along with some carbon. The products evolved are called the volatile matter of the fuel. It is not difficult to determine the temperature at which the release of volatiles ceases if the experiment is done in a pure nitrogen atmosphere. In such a case the accumulative and differential mass loss curves exhibit the typical behavior shown in Figures 5-2 and 5-3 for wood.

a) Almost no release of vapors and gases below 250°C.

b) Rapid devolatilization between 250 and 400°C, the rate reaching its peak at 375°C.

c) Very sharp drop in the release of products after 375°C with termination of the process beyond 450°C.

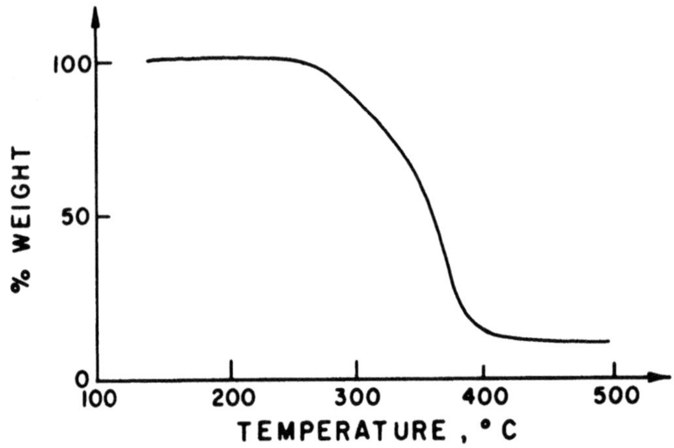

Figure 5-2. Accumulative mass loss curve (modified from Shafizadeh, 1976).

It is noted that in a technical sense pyrolysis and devolatilization are defined as taking place in an oxygen-free atmosphere. Therefore, after the release of all volatiles, carbon can no longer combine in an oxygen-free atmosphere and the process stops completely.

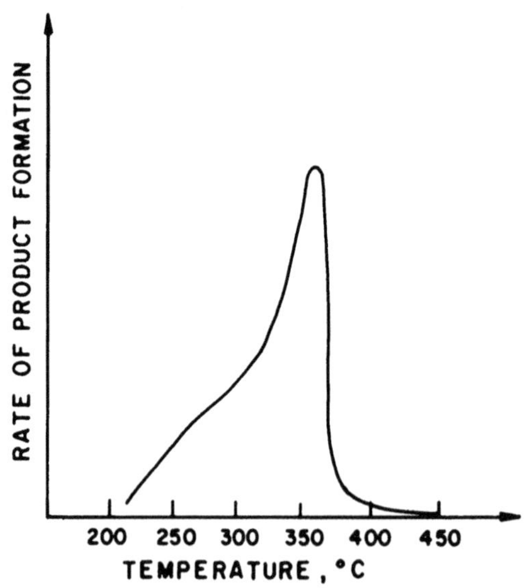

Figure 5-3. Differential mass loss curve. (Shafizadeh, 1976).

The products of pyrolysis of biomass are a mixture of water and condensable higher hydrocarbons, gases, and solid residue. A long list of trace species often found in the condensate of biomass was included earlier (Table 1-1). The proportions of the various products of the pyrolysis of rice hulls at different temperatures will be given later (Chapter 9).

In practice, thermal devolatilization occurs with at least some oxygen present (unless, of course, special steps are taken to exclude it). One of the best known practical applications of devolatilization of biomass is the ancient skill of charcoal production. By burning some of the carbon, the temperatures that are necessary to drive off the volatiles are achieved (around 300°C). It is noted that the devolatilization of biomass is not self-sustaining. The overall heat of reaction oscillates about zero; thus the reaction switches from endothermic to exothermic as shown in Figure 5-4.

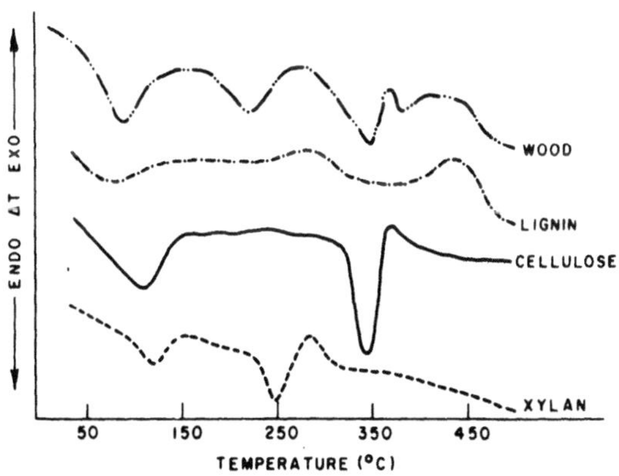

Figure 5-4. Differential thermal analysis. (Shafizadeh, 1976).

A common laboratory method for characterizing biomass fuels is the proximate analysis. In this test the amount of volatile matter, ash and residual carbon in the ash is determined by heating a sample of the biomass to a specific temperature. However, the product distribution is strongly dependent on the temperature, rate of heating and the type of furnace. In addition, the presence of even minute amounts of air can greatly influence the results of these tests. It is therefore not surprising that a wide range of proximate analysis data exist for a given biomass fuel. In general the proximate analysis is not a reliable tool for judging the usefulness of a biomass fuel as a smokeless feedstock for burners and gasifiers because the tests are done in an unspecified atmosphere. Beagle (1978) lists results from previous work. His range for rice hulls are given in Table 5-1. Differences can also occur because of varietal characteristics and processing variables, etc.

Table 5-1. Proximate Analysis of Rice Hulls (% weight, dry basis)

Ash	15.8-23.0
Volatile Matter	56.4-69.3
Fixed Carbon	12.7-17.4

Two widely used methods for pyrolysis analysis of coal are the Fischer and Gray-King Assays. Probstein and Hicks (1982) have carried out a linear least square fit of a large number of standardized 500°C assays for coal. Their work proves that is is possible to predict with high accuracy the fraction of tar + light oils and water released. One interesting part of the evaluation is the linear dependence of the above components on the H/C and O/C ratios. Unfortunately research on the thermal decomposition of biomass has not yet reached the state where a large number of results have been obtained under identical conditions.

The equations for coal are given below and will be compared later to a 500°C assay of rice hulls.

$$\text{Tar + Light Oils (mass\%)} = 29.1 \frac{H}{C} - 12.1 \quad (5\text{-}10)$$

$$H_2O \text{ (mass\%)} = 19.7 \frac{O}{C} + 1.64 \quad (5\text{-}11)$$

One widely used gas producer, the downdraft system, has already been introduced (see Figure 5-1). Another commonly used system, the updraft gas producer, is shown in Figure 5-5. It is noted that no matter how a moving (fixed) fixed bed gas producer is built, there will always be a low temperature zone where pyrolysis takes place, thus a considerable amount of condensable higher hydrocarbons will be generated. In developing a useful model of the gasification process one must exclude this zone and its products because it is very difficult to postulate the products formed, the mode of formation of these products and also the manner in which they are again decomposed.

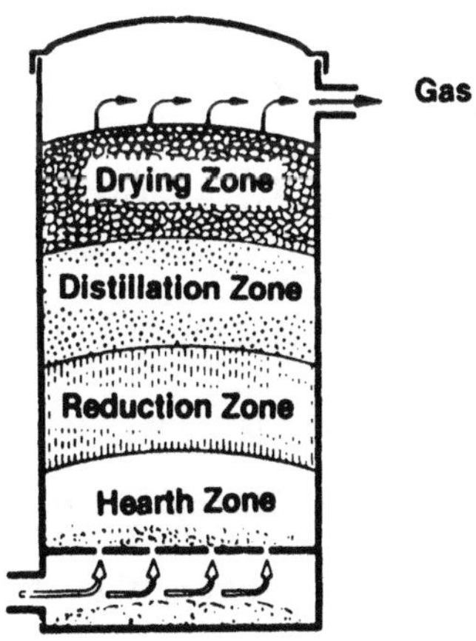

Figure 5-5. Counter-current or updraft gasification (Skov, Papworth, 1975).

It will be shown in Chapter 8 that the mixture of condensable hydrocarbons and H_2O formed is not a small fraction of the pyrolysis products but amounts to over 30% (by weight) of the rice hulls. On the other hand, many gas producers are either operated with devolatilized fuels (charcoal) or at high enough temperatures to induce tar cracking. In these cases, there is no need to account for the products of pyrolysis.

Model I

One of the simplest models which will be called Model I, is based on three assumptions.

Assumption 1:

a) The two input streams, fuel and air, are given as (see 5.1):

Fuel: $C_{3.45}\,H_{4.94}\,O_{2.33}\,(SiO_2)_{0.26}$ (100 kg of dry rice hulls)

Air: $N_{1.58}\,O_{0.42}$ (1 kg-mole air)

For an equivalence ratio of $\phi = 1$ the chemical composition of the two feed streams is:

$$C_{3.45}\,H_{4.94}\,O_{9.38}\,N_{26.52}\,(SiO_2)_{0.26}$$

It is noted that Silicon (Si) is already in its oxidized form in the rice hulls and will not be oxidized by the combustion air. Silicon is most often found in plants within a Silicate System. The representation as SiO_2 only serves to indicate that there will be no oxygen consumed by this mineral in rice hulls. With biomass fuels having an ash content smaller than 1% such as corn cobs and most woods, the ash may be neglected. However as shown in Figure 8-11, the ash content of rice hulls is too high to neglect in computations.

At first, one may find it strange to represent a mixture of solid carbonaceous fuel and air in the form of atomic ratios which does not reveal the actual nature (gas, liquid, solid) of the inputs. However, the product concentrations are functions only of temperature and pressure and not the original arrangements of the atoms. Thus, for a computer program "combusting" a given fuel, it does not matter whether the original atoms are in a gaseous, solid or liquid form.

Assumption 2:

All possible reversible reactions go to equilibrium. This assumption is not completely valid for two reasons:

a) The pyrolysis process is highly irreversible and there is a wide range of intrinsic gasification rates for the Equations 5-1 through 5-5. How the intrinsic rates compare is given by Walker et al. (1959) and listed in Table 5-2.

Table 5-2. Relative Intrinsic Gasification Rates at 1 atm and 800°C

Reaction	Relative Rate
$C + O_2 \rightleftarrows CO_2$	10^5
$C + H_2O \rightleftarrows CO + H_2$	3
$C + CO_2 \rightleftarrows 2\ CO$	1
$C + 2\ H_2 \rightleftarrows CH_4$	10^{-3}

It is noted that the water shift reaction (Equation 5-3) is the only significant homogeneous reaction taking place. The assumption of equilibrium is therefore justified but nevertheless difficult to prove for several reasons:

a) In an updraft gas producer the off gas is always mixed with the products (gases, tar and water) of the pyrolysis zone. Any reliable verification of the water shift reaction reaching equilibrium must be done below the pyrolysis zone. In addition a "true" temperature must be measured.

b) The same difficulties arise with the off gas from a downdraft gas producer. In a well-functioning unit, no uncracked pyrolysis products, except water, will be present in the raw gas. However, one still faces the difficulties involved in determining a "true" temperature for the gas sample.

To verify whether the equilibrium state of the water shift reaction is reached, it is best to use a fluidized bed gas producer, because a more homogeneous temperature and species concentration field is present. In fact, by definition the species concentrations and temperature field are independent of the spatial coordinates in a fluidized bed gas producer.

The degree to which the water shift reaction approaches equilibrium in such a unit is shown in Figure 5-6. The conversion efficiency, η, is represented by:

$$\eta = \frac{(CO_2)_{max} - (CO_2)_{meas}}{(CO_2)_{max} - (CO_2)_{equil}}$$

The graph indicates that the product distribution is dominated by pyrolysis products in the low temperature range. In the high temperature range, the graph shows that the water shift reaction is near completion. Based on these results it is of interest to check the temperature sensitivity of the equilibrium constant, K_p (T).

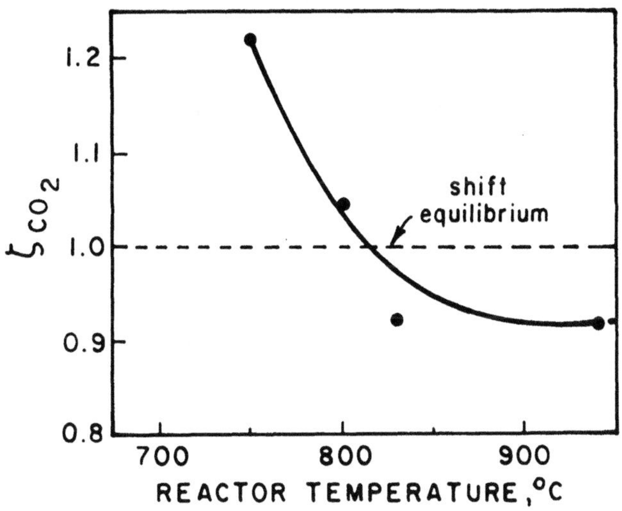

Figure 5-6. Equilibrium of the water shift reaction as a function of temperature in a fluidized bed rice hull gasifier (Van den Aarsen et al., 1982).

K_p (T) is given by:

$$K_p(T) = e^{-\frac{\Delta^\circ G(T)}{RT}} = \frac{[CO_2]\,[H_2]}{[CO]\,[H_2O]} \quad (5\text{-}12)$$

K_p (T) as a function of T is shown in Figure 5-7. The standard free energy change for the reaction when going to completion at 1 atm is given by $\Delta^\circ G(T)$. Its value can be computed in KJ/g-mol by:

$$\Delta_i G^\circ(T) = \Delta^\circ{}_i H_{298} - a_i T \ln T - b_i T^2 - \frac{c_i}{2} T^3 - \frac{d_i}{3} T^4 + \frac{e_i}{2}\frac{1}{T} + f_i + g_i T \quad (5\text{-}13)$$

where i represents the species CO, CO_2, H_2O and H_2. Then $\Delta^\circ G(T)$ is given by:

$$\Delta^\circ G(T) = \sum_{i=1}^{n} \Delta_i^\circ G(T) \qquad (5\text{-}14)$$

The coefficients in Equation (5-13) are tabulated by Probstein and Hicks (1982).

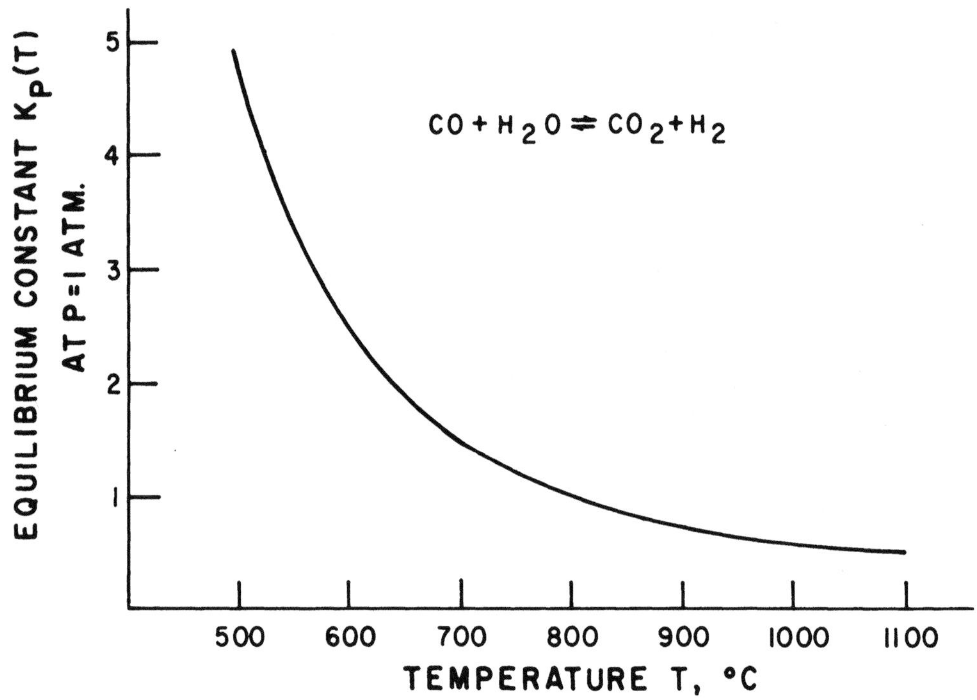

Figure 5-7. K_p (T) as a function of T.

Example

The following example shows that it is not always possible to determine the validity of Assumption 2.

A typical raw gas composition as measured in one of the small rice hull downdraft gas producers is the following:

Species	% Volume
CO_2	12.6
CO	15.3
H_2	10.8
H_2O	10.1

The associated equilibrium temperature is 840°C. The measured temperature was 742°C. At 742°C, the equilibrium composition should be:

Species	% Volume
CO_2	13.4
CO	14.5
H_2	11.6
H_2O	9.3

From the example it is clear that the water shift reaction due to its small energy change with respect to large temperature changes is very stable. The vastly different response times of the gas analyser combined with the inaccuracy of high temperature measurements in a convective gas stream prohibits any firm conclusions from pilot plants.

Assumption 3

No condensable products except for water, minerals and solid carbon are present in the product stream.

Basically this means a "tar-free" gas, thus no carbon or hydrogen is bound in higher hydrocarbons (> C_5H_m). This assumption cannot be valid for most gasifiers. It may be justified for well-designed downdraft gas producers.

Computer Program

Based on these three assumptions, the raw gas compositions of a rice hull "reactor" operating at various equivalence ratios have been computed. Reactors operating without an external heat source are sometimes labeled gas producers whenever the gas contains enough chemical energy to allow subsequent combustion in a furnace or internal combustion engine.

The computations have been done with the STANJAN program developed by R. Reynolds (1982). This program was selected over the other available programs for the following reasons:

a) The program uses one of the largest thermodynamic data bases presently available (JANAF Tables, 1971).

b) It allows the user to specify the products. For the purposes of this work, the following product species were chosen:

Solids: C

Liquids: H_2O (ℓ)

Gases: CH_3OH, CH_4, C_2H_2, C_2H_4, C_2H_6, H, H_2, H_2O(g), HO, N, N_2, 25, NO_2, O, O_2, C(g).

c) The program accounts for condensable phases and for the products of disassociation reactions which become important at higher temperatures.

The results of the computations are given in Figures 5-9 through 5-22. Species having mole fractions less than 0.01 are not shown in the graphs. It is noted that the gas composition has been computed over a temperature range which will not be attainable without external heating for small values of ϕ. For instance, sustaining the gasification process at ϕ = 0.1 and T = 1000°C is hardly possible, because the heat generated from the small amounts of carbon burned is not enough to match the heat losses from the reducing reactions and from the system (see Section 9.4).

Another limitation lies in Assumption 3. It will be shown in Chapter 9 that insignificant tar conversion occurs at temperatures below 800°C. The graphs are therefore only realistic within a narrow temperature band with a lower limit of 800°C. The upper limit will be the adiabatic combustion temperature at a given ϕ.

Let c be the raw gas composition, then

$$c = c(\phi, M, T)$$

where: ϕ = equivalence ratio

M = moisture content in % weight, wet basis

T = temperature, °C.

The range of cases shown in Figures 5-9 through 5-21 for the temperature interval 500°C to 1400°C is shown in the grid in Figure 5-8.

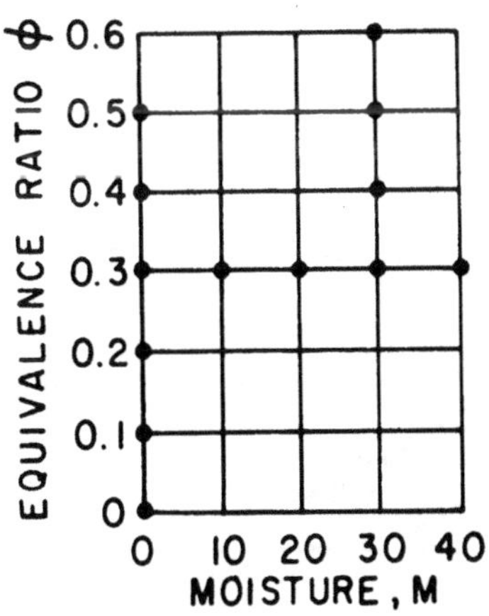

Figure 5-8. Range of computed gas compositions as a function of M and ϕ.

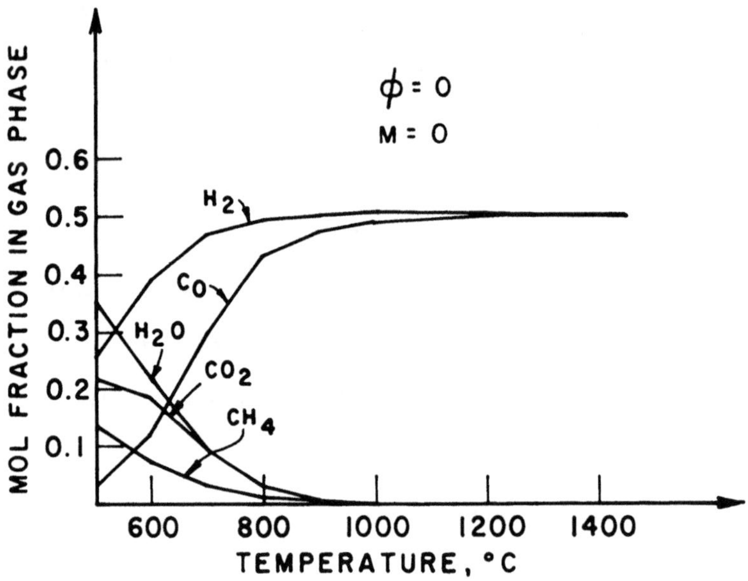

Figure 5-9. Gas composition, ϕ = 0, M = 0

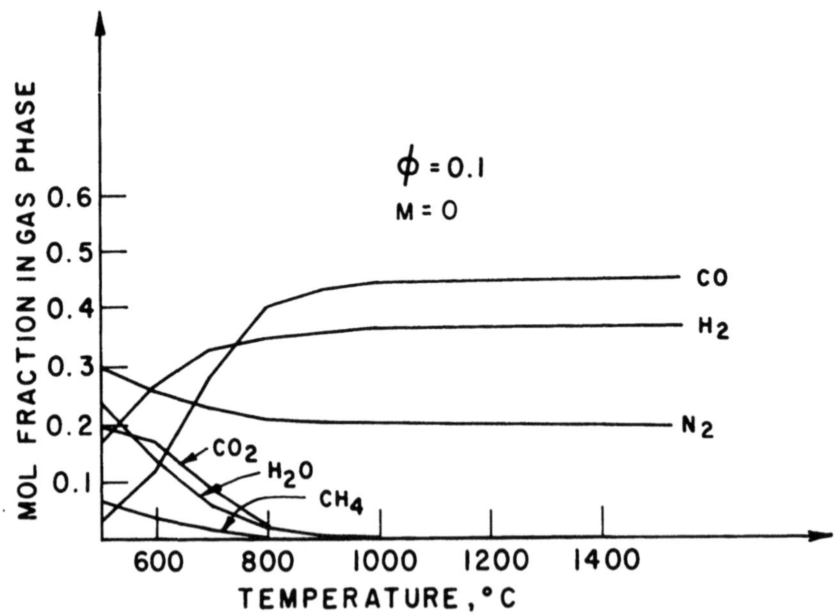

Figure 5-10. Gas composition, ϕ = 0.1, M = 0

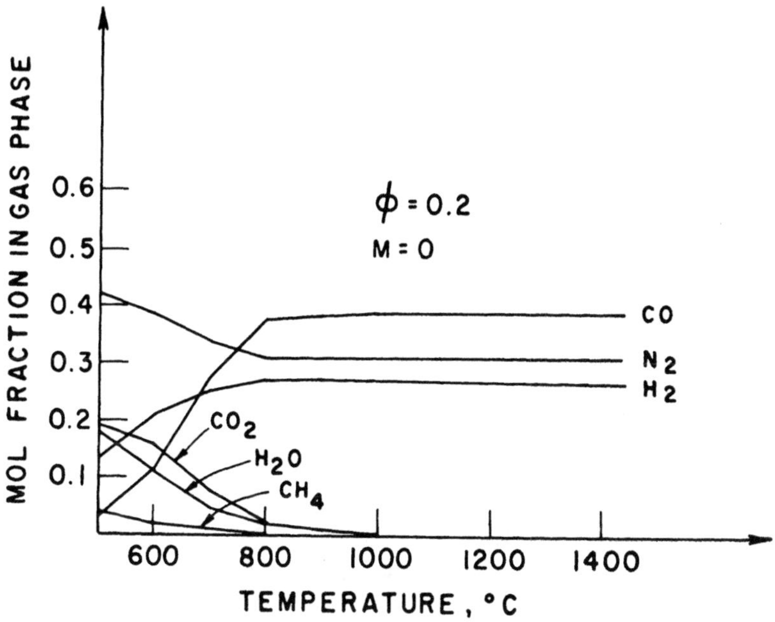

Figure 5-11. Gas composition, ϕ = 0.2, M = 0

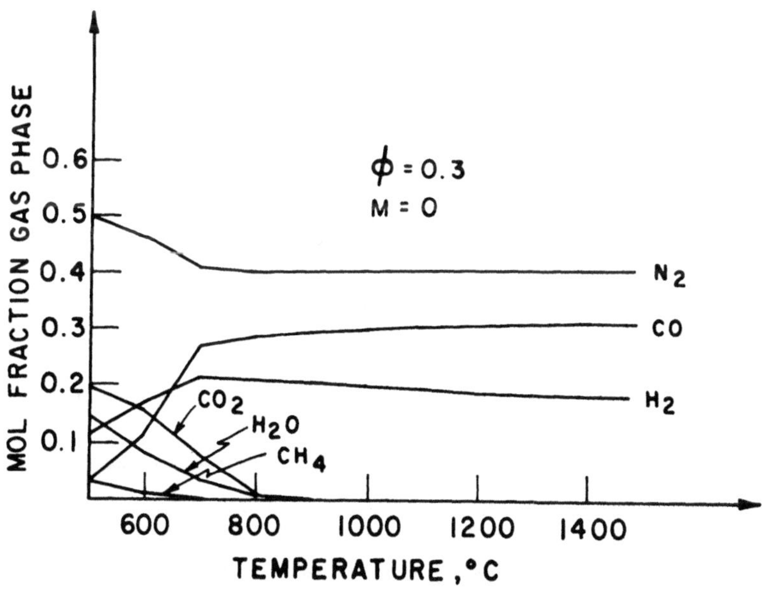

Figure 5-12. Gas composition, ϕ = 0.3, M = 0

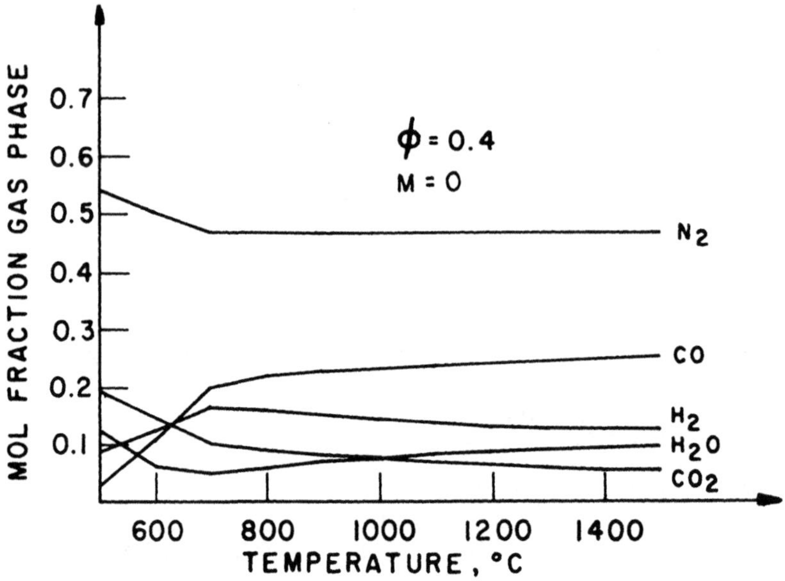

Figure 5-13. Gas composition, ϕ = 0.4, M = 0

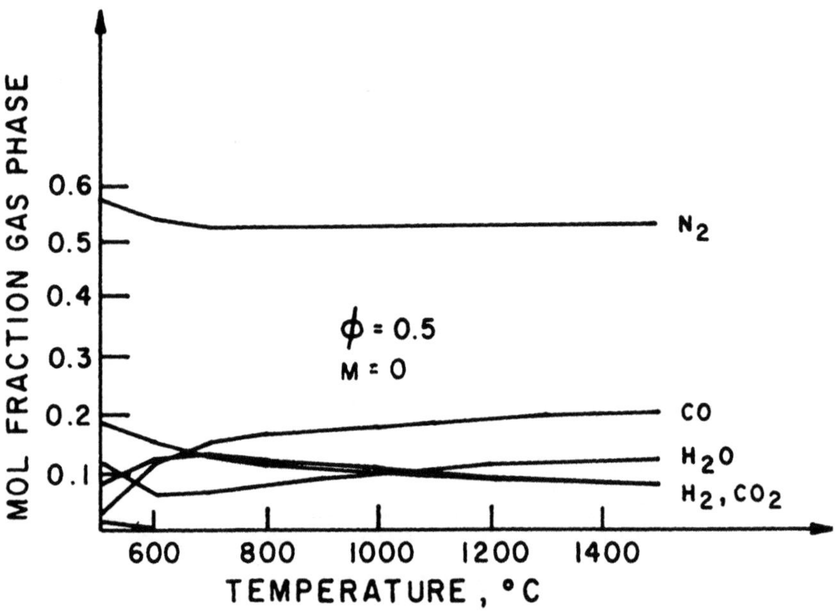

Figure 5-14. Gas composition, ϕ = 0.5, M = 0

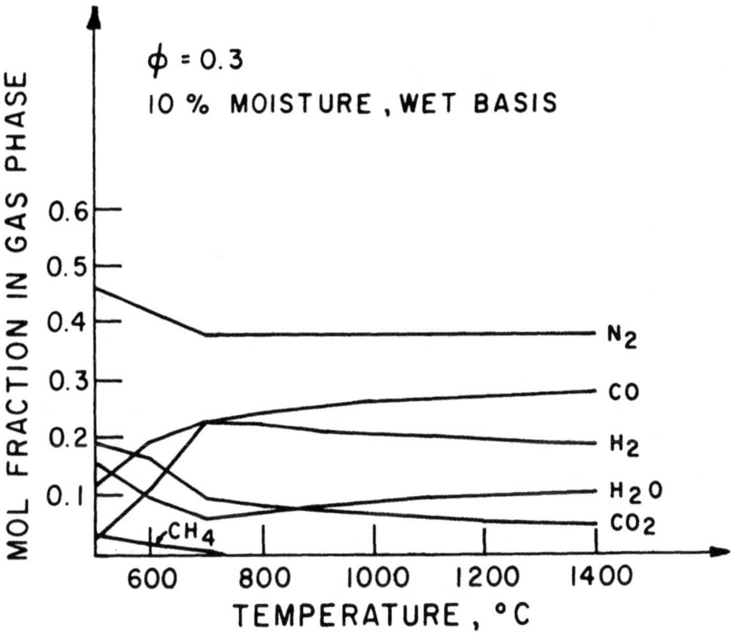

Figure 5-15. Gas composition, ϕ = 0.3, M = 10

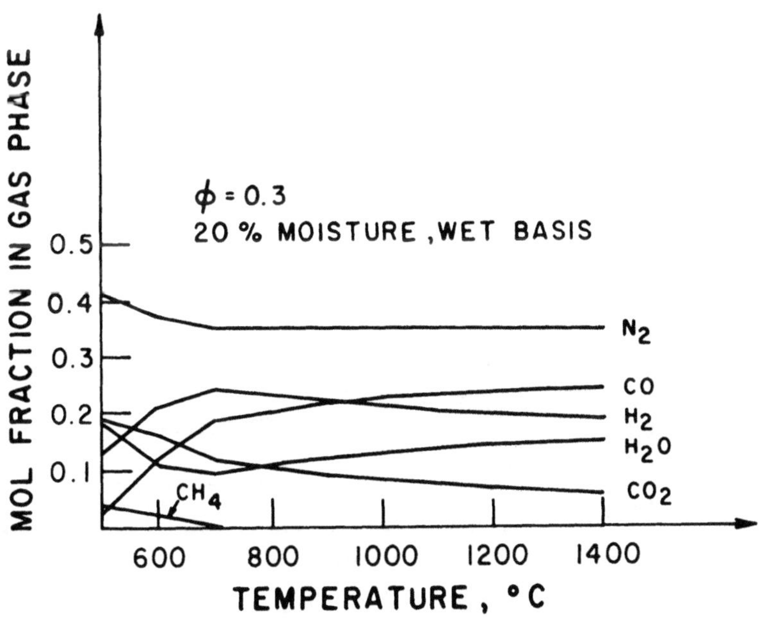

Figure 5-16. Gas composition, ϕ = 0.3, M = 20

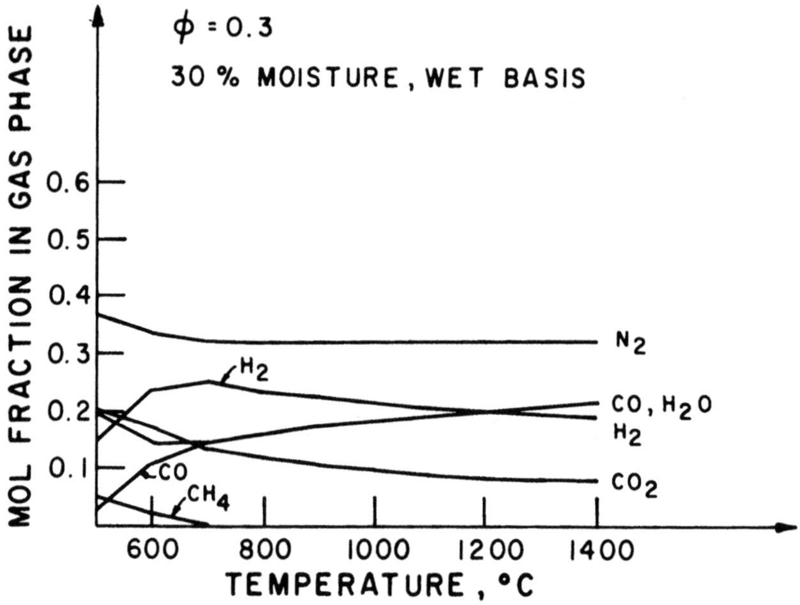

Figure 5-17. Gas composition, $\phi = 0.3$, M = 30

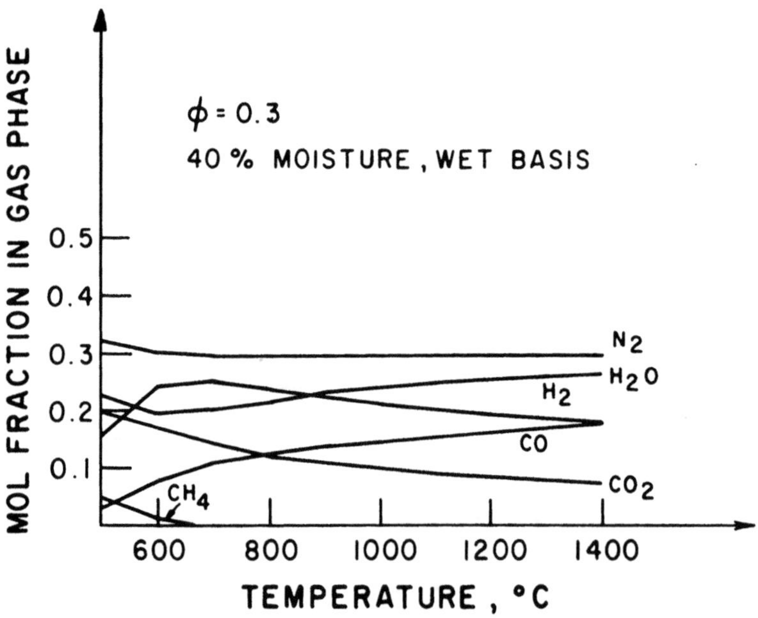

Figure 5-18. Gas composition, $\phi = 0.3$, M = 40

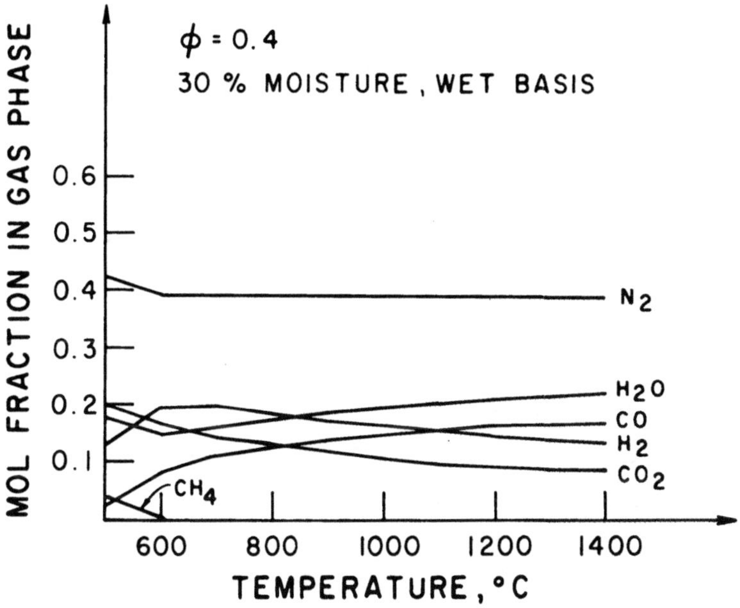

Figure 5-19. Gas composition, ϕ = 0.4, M = 30

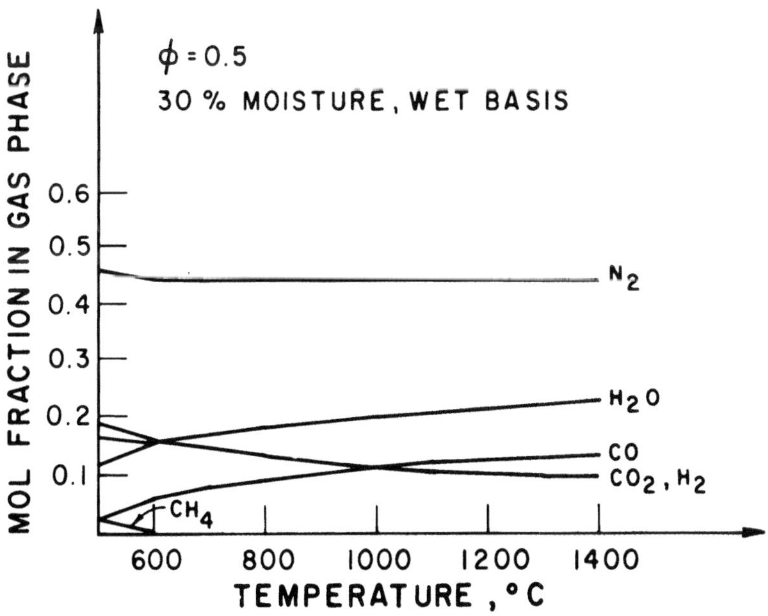

Figure 5-20. Gas composition, ϕ = 0.5, M = 30

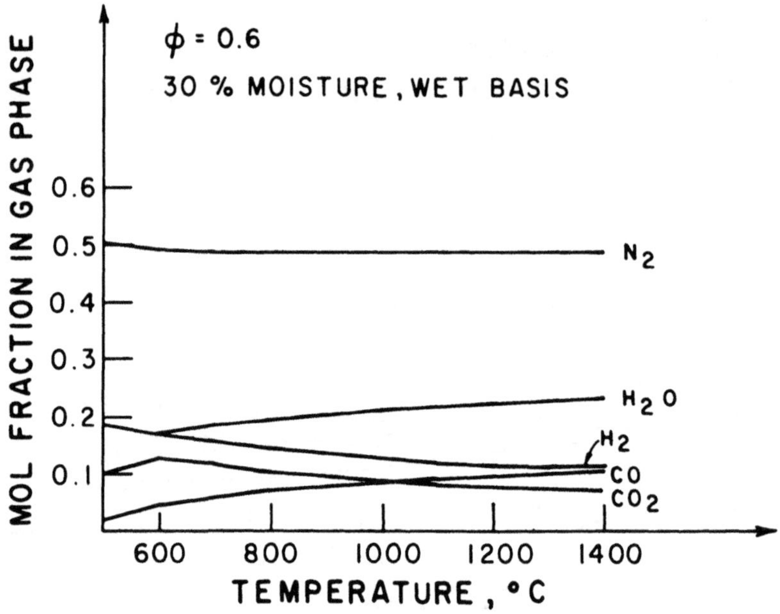

Figure 5-21. Gas composition, ϕ = 0.6, M = 30

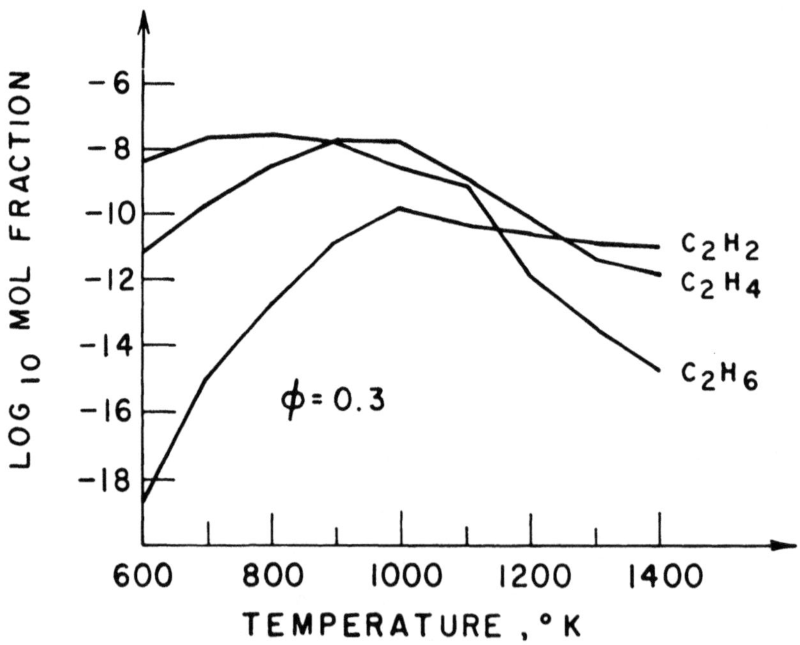

Figure 5-22. C_2H_m equilibrium composition at various temperatures, ϕ = 0.3

Analysis of the Results

Although no mass transport, kinetics and local energy balances are incorporated in Model I some important conclusions can be drawn.

a) Hydrocarbons CH_4, C_2H_m, C_3H_m are not formed from molecular or atomic hydrogen present in the gas phase. At temperatures above 800°C none of these hydrocarbons are formed in significant amounts. (See Figure 5-22). Consequently, large quantities of gaseous total hydrocarbons in the sample gas are an indicator that the unit is generating a highly contaminated raw gas and operating at a low temperature.

b) Operating a reactor at a low equivalence ratio results in high CO and H_2 fractions at high temperatures. However in practice such conditions are not possible without external heating. This last statement has important implications with regard to the design of a gas producer and with regard to the condition (moisture) of the rice hulls.

In order to achieve the favorable combination of low ϕ and high temperatures it is necessary to:

(i) Heavily insulate the gas producer.

(ii) Preheat the combustion air with the sensible heat of the raw gas, using a heat exchanger.

(iii) Use rice hulls that are as dry as possible.

Influence of the Moisture Content

The moisture content of rice hulls depends on handling and milling procedures (commercially 8-12%) as well as the local storage and climate. Moisture in the feed material has a marked effect on the raw gas composition and efficiency of the process. There are significant differences between fluidized bed gasifiers, updraft gasifiers and downdraft gasifiers with respect to their ability to cope with high moisture contents of the rice hulls when on an experimental basis.

In an updraft gas producer the moisture in the fuel does not come into contact with a hot carbon zone (> 800°C). Therefore the only detrimental effect of a high-moisture content is a reduction in efficiency, because more carbon must be combusted to provide the necessary heat for evaporation of the water. In addition, the gas stream loses a fraction of its sensible heat, in heating up the wet feed material.

In downdraft and fluidized bed gas producers with bottom feed, the water evaporated from the fuel must pass through high temperature zones rich in carbon, where it is converted to a certain extent to water gas. It will be shown in Section 9.4 that the water gas reaction (5-2) is slow and not very effective in a rice hull fuel bed.

The effect of the moisture content on the gasification reactions and product composition is shown in Figures 5-13 through 5-21. The first set of graphs (5-15 through 5-18) shows an operation at ϕ = 0.3 with rice hulls having a 10%-40% moisture content. It is noted that the water vapor content of the gas increases significantly at the expense of CO with insignificant change in the hydrogen content of the raw gas. The trend is shown for one specific temperature (T = 1000°C) in Table 5-3.

Table 5-3. Species Concentration at T = 1000°C and ϕ = 0.3

Species	% volume at various M			
	10%	20%	30%	40%
CO	26.1	22.3	18.6	15.0
CO_2	6.6	8.2	9.5	10.3
H_2	20.6	21.2	21.5	21.2
H_2O	8.6	12.8	18.0	24.0
CH_4	0	0	0	0

In practice, an increasing moisture content necessitates an increasing ϕ in order to provide the necessary heat of reaction to reach reasonable gasification temperatures (1000°C and higher). The heating value of the gas is greatly decreased. (See Figures 5-19 through 5-21). In this set of figures the equivalence ratio ϕ is varied using a fixed moisture content of 30%. As can be seen from Table 5-4 the raw gas composition shows a sharp decrease in the combustible fractions of CO and H_2 with increasing ϕ, a corresponding increase in the sensible heat will take place.

Table 5-4. Species Concentration at T = 1000°C, M = 30%, wet basis

Species	% volume at various ϕ			
	0.3	0.4	0.5	0.6
CO	18.6	14.7	11.4	8.6
CO_2	9.5	10.6	11.7	12.6
H_2	21.5	16.3	12.1	8.7
H_2O	18.0	19.4	20.4	21.1
CH_4	0	0	0	0

This behavior of the gasification reactions is one of the most often overlooked disadvantages of gasifiers that are capable of cracking tar and decomposing moisture. The high temperatures and significant energy needed for the conversion of both tar and water can only be achieved by drastically increasing the equivalence ratio for biomass having a high moisture content. This in turn leads to internal burning of the combustible fractions, CO and H_2, and a shift from the chemical energy to sensible heat.

Higher Heating Values of the Computed Gas Compositions

The higher heating values of the raw gas composition are given in Figure 5-23. The heating value decreases significantly with increasing ϕ, because of the shift from chemical energy to sensible heat as shown in Figure 1-1. However, a more interesting feature of the raw gas composition is its nearly constant heating value at temperatures above 900°C, for all ϕ. The reason for this phenomenon is that the water shift reaction dominates at higher temperatures causing only a shift in the gas composition towards a higher CO/H_2 ratio. The water shift reaction has little effect on the total chemical energy since H_2 and CO have about the same higher heating value (286 MJ/kg-mol for H_2 and 283 MJ/kg-mol for CO, at 25°C and 1 atm). It is noted that operation at higher temperatures than what is required to crack tar will therefore not result in a gas with a higher heating value.

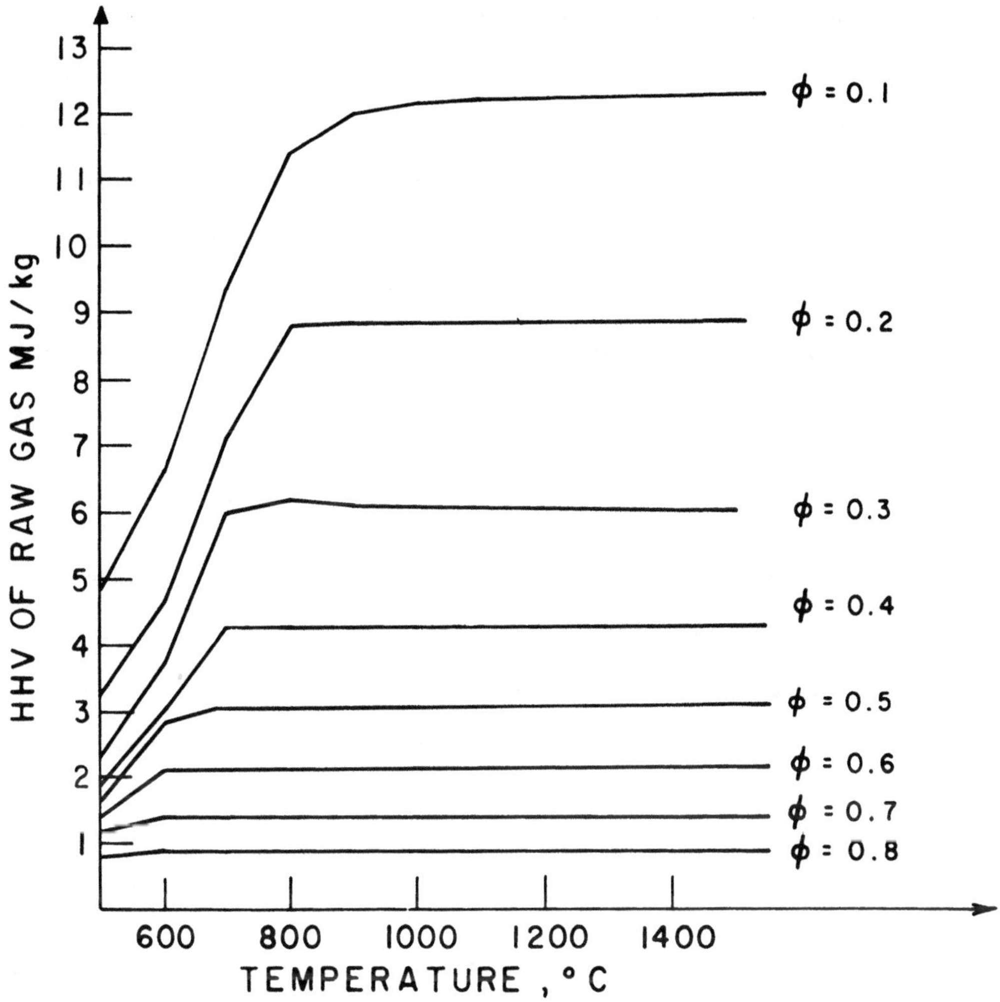

Figure 5-23. Higher heating value of the raw gas for range of T and ϕ.

Summary for Model I

a) The raw gas composition was computed as functions

$$c_i = c_i\ (\phi,\ M,\ T),\ i = 1,2,3,N.$$

b) The results apply only to gasifier designs that allow the fuel moisture and pyrolysis products to be converted into gaseous products.

c) The computed gas composition must be viewed as the best that can be achieved at any given temperature, irrespective of whether this temperature is obtainable with the given ϕ. One exception is the methane content of the gas. It is usually higner in experimental data, since total tar conversion rarely takes place.

d) It will be shown later (Section 9.4) that the steam conversion is far from completion at any temperature in a rice hull fuel bed with its high void space fraction. The actual water content of the gas is therefore higher than the computed one.

e) The gasification of rice hulls with a very high moisture content (> 30%) seems to be impractical in downdraft units and fluidized beds. The issue is not the decreasing efficiency of the process but the high ϕ needed to sustain temperatures favorable for tar cracking and moisture conversion.

f) Hydrocarbon gases in the product stream are unconverted pyrolysis products and indicate low temperature gasification.

It is noted that this simple concept of gasification, viewing the reactor as a black box with a known input stream, does not account for mass transport phenomena, kinetic rates or local energy balances. However the advantage of Model I is that it is easily solvable and is very helpful in pointing out general trends. Taking into account intrinsic kinetic rates of the various reactions and mass transport limitations, Groeneveld (1981) in his dissertation, compared more sophisticated models to his experimental data from a small charcoal and wood downdraft gas producer.

One could also follow a completely different path, as outlined by Whitaker (1981) and argue that since the process must occur according to the basic laws of mechanics, it is always possible to write down the "exact" governing differential equations for the process.

Model II

The difficulties with Model II are the following:

a) The "exact" equations express all variables and physical parameters as a function of the spatial coordinates and time.

b) Boundary conditions, which depend heavily on the design of the gas producers, must be established.

c) Any measurements taken constitute an average bulk phase result; it is therefore necessary to volume average the point equations in order to compare theory and experimental data.

d) The "exact" differential equations must be modified to a point where they can be expressed in terms of physical parameters that are readily measurable such as temperature, mass velocity fields, thermal conductivity, void space fractions and diffusion coefficients.

e) An order of magnitude analysis must be performed to exclude all terms that do not contribute significantly to the process.

f) All processes take place within a three-phase system - 1. void space in the bed; 2. solid phase in rice hulls; 3. gas phase in rice hulls.

It is noted that the limitations of this method are not due to a lack of computer programs to perform the necesary computations, but are related to the lack of knowledge concerning the magnitudes of the physical parameters and their relative importance.

Although it is beyond the scope of this work to outline in detail the complications involved in establishing a set of solvable equations a short outline is given. The species continuity equation is given as

$$\frac{\partial c_i}{\partial t} + \nabla \cdot (c_i \, v_i) = R_i \tag{5-15}$$

The energy equation will be represented by

$$\rho c_p \frac{DT}{Dt} = - \nabla \cdot q - \nabla \cdot \Sigma \, J_i \, \bar{H}_i + T\beta \, \frac{Dp}{Dt}$$

$$+ \nabla \, v : t + \Phi - \sum_{i=1}^{n} \bar{H}_i \, R_i + \sum_{i=1}^{n} \bar{H}_i \, \nabla \cdot J_i \tag{5-16}$$

It is noted that:

a) The index, i, stands for the various major gas species, CO, CO_2, H_2, H_2O, N_2, O_2, CH_4.

b) Both equations are coupled by an unknown velocity field which can be determined by the Navier Stokes equations.

c) The displacement of a gas molecule by diffusion (caused by concentration gradients) is ignored because it is insignificant when compared to convective displacement.

d) Based on typical designs of rice hull gas producers no variations of c_i in the radial direction is assumed and therefore only concentration gradients in the axial direction must be taken into consideration.

Species Concentration Equation

As already mentioned it is necessary to convert the unworkable point equation into a volume averaged differential equation. This is done by first defining the size of the control volume V in which physical and chemical properties are considered constant. The general structure of a rice hull control volume is shown in Figures 6-1 and 6-2. The various phases and numerical values of their fractions have been experimentally determined and are reported in Chapter 6. We define for the β-phase:

$$\langle c_i \rangle = \frac{1}{V} \int_{V_\beta} c_i \, dV \tag{5-17}$$

$$\langle f \rangle = \frac{1}{V} \int_{V_\beta} f \, dV \tag{5-18}$$

for any parameter f we consider constant within the control volume V. For purely technical reasons one may wish to define

$$\langle f \rangle^\beta = \frac{1}{V_\beta} \int_{V_\beta} f \, dV$$

The relationship between $\langle f \rangle^\beta$ and $\langle f \rangle$ is given by $\langle f \rangle = \varepsilon_\beta \langle f \rangle^\beta$ where ε_β represents the void space fraction given as

$$\varepsilon_\beta = \frac{V_\beta}{V} \tag{5-19}$$

The process of volume averaging then leads to

$$\nabla \langle \mathbf{v}_i \, c_i \rangle + \frac{1}{V} \int_{A_{\beta\sigma}} c_i \, (\mathbf{v}_i - \mathbf{w}) \cdot \mathbf{n}_{\beta\sigma} \, dA = \langle R_i \rangle \tag{5-20}$$

It is noted that $\langle R_i \rangle$ is the "production" rate of species i in the gas phase. This term is non-zero only for the homogeneous water-shift reaction. All other reactions are assumed to be heterogeneous. They are accounted for by the integral term

$$\int_{A_{\beta\sigma}} c_i \, (\mathbf{v}_i - \mathbf{w}) \cdot \mathbf{n}_{\beta\sigma} \, dA$$

which describes the mass flux (mole m^{-2} s^{-1}) of species i across the apparent surface $A_{\beta\sigma}$ as shown in Figure 6-2. The velocity **w** is the velocity in the interphase.

It is, of course, up to the individual to decide which model will best fit this interfacial mass transport. It is noted from Figure 6-2 and the micrographs 7-14 and 7-15 that the interfacial area $A_{\beta\sigma}$ is the sum of the areas associated with the solid carbon phase and the pore entrances. The rice hull velocity in the reactor is assumed to be identical with the average of **w**.

Comparing micrograph Figures 7-4 and 7-12 one can see that rice hulls exhibit a rather unusual behavior during thermal decomposition. Their outer surface appears to be impermeable both before and after gasification. In fact, no visible pores could be detected at a magnification of 6000x. On the other hand, the inner surface becomes extremely porous. If one assumes that no appreciable surface reactions occur on the internal pore surfaces, then a manageable representation of the interfacial mass flux can be developed. Based on Equations (5-1) through (5-5) we can assume the following sequence of events:

a) A gaseous species i (for instance CO_2) is diffusing through a boundary layer towards the carbon surface.

b) CO_2 is adsorbed by the surface at a specific rate $k_{ad} \cdot C_{CO_2}$.

c) CO_2 will be present at the carbon surface at a concentration of $\widehat{C}_a$.

d) CO_2 will react according to $C + CO_2 \rightleftarrows 2\ CO$.

e) The product CO will be desorbed and will diffuse back into the gas phase.

The missing link is the constituent equations for c) and d). For instance, if one assumes first order irreversible surface kinetics and a surface concentration proportional to the bulk phase concentration c_i we have

$$c_i\ (\mathbf{v}_i - \mathbf{w}) \cdot \mathbf{n}_{\beta\sigma} = \kappa\hat{k}\ c_i \tag{5-21}$$

at the surface and the integral term is now modelled by

$$\frac{1}{V} \int_{A_{\beta\sigma}} c_i\ (\mathbf{v}_i - \mathbf{w}) \cdot \mathbf{n}_{\beta\sigma} = a_{\beta\sigma}\ \kappa\hat{k}\ \langle c_i \rangle_{\beta\sigma} \tag{5-22}$$

with
$$\langle c_i \rangle_{\beta\sigma} = \frac{1}{A_{\beta\sigma}} \int_{A_{\beta\sigma}} c_i\ dA \tag{5-23}$$

and $a_{\beta\sigma}$, the interfacial specific area (m^2/m^3) that has been measured and is given in Chapter 6. Equation (5-21) now reduces to

$$\nabla \cdot \langle \mathbf{v}_i c_i \rangle + a_{\beta\sigma}\ \kappa\hat{k}\ \langle c_a \rangle_{\beta\sigma} = \langle R_i \rangle \tag{5-24}$$

Since $\langle \mathbf{v}_i c_i \rangle$ is not measurable it is necessary to express it in terms of known parameters.

Let
$$\mathbf{v}_i = \langle v_i \rangle^{\beta} + \tilde{\mathbf{v}}_i \tag{5-25}$$

$$c_i = \langle c_i \rangle^\beta + \tilde{c}_i \tag{5-26}$$

Thus, the point concentration and velocity of species i are split up into the sum of the measurable average bulk phase values and a spatial deviation. It is not unreasonable to assume that $\langle \tilde{v}_i \rangle = \langle \tilde{c}_i \rangle = 0$. In fact one simply chooses a very small control volume so that the deviations integrated over the volume cancel one another. Applying the volume averaging rule to equations (5-24) through (5-26) we have

$$\nabla \varepsilon_\beta (\langle v_i \rangle^\beta \langle c_i \rangle^\beta) + \nabla \langle \tilde{v}_i \tilde{c}_i \rangle = \langle R_i \rangle - a_{\beta\sigma} \kappa\hat{k} \langle c_a \rangle_{\beta\sigma} \tag{5-27}$$

Based on theoretical considerations (Carbonell, Whitaker, 1983) one may let

$$\langle \tilde{v}_i \tilde{c}_i \rangle = -\varepsilon_\beta D \nabla \langle c_i \rangle^\beta \tag{5-28}$$

where D is the hydrodynamic dispersion tensor. We therefore have:

$$\nabla (\varepsilon_\beta \langle v_i \rangle^\beta \langle c_i \rangle^\beta) = \nabla (\varepsilon_\beta D \nabla \langle c_i \rangle^\beta) + \varepsilon_\beta \langle R_i \rangle^\beta - a_{\beta\sigma} \kappa\hat{k} \langle c_i \rangle_{\beta\sigma} \tag{5-29}$$

Gasification of rice hulls can be described as a radial symmetric process. It was previously assumed that D_{rr} is zero and that no convective transport in the radial direction takes place. This leads to

$$\frac{d}{dz} (\varepsilon_\beta \langle v_z \rangle^\beta \langle c_i \rangle^\beta) = \frac{d}{dz} \left(\varepsilon_\beta D_{zz} \frac{\partial \langle c_i \rangle^\beta}{\partial z} \right) + \varepsilon_\beta \langle R_i \rangle^\beta - a_{\beta\sigma} \kappa\hat{k} \langle c_i \rangle_{\beta\sigma} \tag{5-30}$$

where $v_i = v_z$ is assumed.

It is noted that D_{zz} refers to the axial dispersion coefficient for species i in a mixture. From equation (5-30) it can be seen that the solution of this fairly simple differential equation contains several physical parameters such as

a) Void space fraction, ε_β

b) Dispersion coefficient, D_{zz}

c) Specific surface area, $a_{\beta\sigma}$

d) Linear adsorption isotherm, κ

e) Surface reaction rate, $\hat{k}$

These parameters must be determined and are, of course, functions of z although some of them, such as ε_β, can be assumed to be constant (see Chapter 6). In addition, the velocity field must be established. The bulk phase velocity $\langle v_z \rangle$ is certainly a function of z because the volumetric gas output is larger than the air input and the density of the gas mixture is proportional to the temperature ratio. The horizontal velocity field in small rice hull gas producers is unusual, as will be explained later (see Section 6.6). In the case where there are homogeneous bulk phase reactions (water shift reaction) a constitutive equation for $\langle R_i \rangle^\beta$ must be found. Because of the highly endothermic nature of the water gas and CO_2-reduction reactions the term $\langle R_i \rangle^\beta$ will be a function of z.

In summation: At this point the solution to the "exact" set of differential equations can not be found due to the lack of data concerning physical parameters. It is noted that this type of approach may also be used to derive a set of equations that take into account the influence of the internal pores of rice hulls. In addition, the above mentioned axial dispersion coefficient D_{zz} does not include turbulent effects, which may be overwhelming.

The Energy Equation

The complete equation (5-16) cannot be solved in this form. A check of the dimensions shows that each term represents the rate of change of internal energy per unit control volume (J m^{-3} s^{-1}). However one can show that the following terms contribute insignificantly and will be neglected:

a) $\sum_{i=1}^{n} H_i \nabla \cdot \mathbf{J}_i$ and $\sum_{i=1}^{n} J_i \overline{H}_i$ which describe the energy transport by diffusion of the individual species and are small.

b) The reversible component of the work term $T\beta\ \mathbf{v} \cdot \nabla p$ which accounts for the compression and expansion of the gases is neglected.

c) The irreversible component of the work term $\nabla \cdot \mathbf{v} : \mathbf{t}$ which accounts for the heat dissipation due to friction will also be neglected.

In summation, the overall energy equation can be greatly simplified. The result is the following:

$$\rho c_p\ \mathbf{v} \cdot \nabla T = -\nabla \cdot q + \Phi - \sum_{i=1}^{n} \overline{H}_i\ R_i \tag{5-31}$$

The term Φ represents an internal heat source, which describes radiation. Radiation can not be neglected at the temperatures that prevail in gas producers. The energy change due to the highly endothermic and exothermic reactions taking place in a gas producer are represented by:

$$\sum_{i=1}^{n} \overline{H}_i R_i \qquad (5\text{-}32)$$

where R_i is the molar rate of "production" of the i-th species and H_i is its partial enthalpy (J/mole mixture). It is noted that the latter term includes the heat of formation and not just the sensible heat change. It is therefore given by Equation (5-33).

At this point, further simplification is only possible if the effects of radiation within a packed bed can be assessed. It has been concluded that an exact description of the rice hull gasification process by the previously outlined method must be preceded by a more detailed analysis of the physical and kinetic parameters involved.

Flame Temperatures of Producer Gas

Introduction

Equilibrium compositions for a wide range of equivalence ratios, rice hull moisture contents and temperatures have been computed in Section 5.2. Once the equilibrium composition is known, one can then examine the possible adiabatic flame temperatures and the parameters which influence these temperatures.

Using producer gas for external combustion (Stirling Engine) or direct combustion (furnace, boiler applications) is less complex than using producer gas in an internal combustion engine, because of the extensive cleaning of the gas which must take place before the gas can be used in an I.C. engine. Some applications such as crop drying can require a heat exchanger because the crop may otherwise be contaminated (odor, particles) by the products of combustion. It is therefore important to anticipate the various maximum temperatures that may occur. The following cases have been studied:

a) Flame temperature as a function of the equivalence ratio, A, for combustion of producer gas.

b) Flame temperature as a function of the products of combustion.

c) Flame temperature as a function of the mixture inlet temperature.

d) Flame temperature as a function of the solid carbon content of the raw gas.

e) Comparison with reported experimental data.

Computer Program

Let g be the adiabatic flame temperature; g then represents a function g(ϕ, T, M, A) in c where:

T = Temperature of equilibrium gas composition, °C

M = Moisture content of the fuel, % weight, wet basis

ϕ = Equivalence ratio for the gasification process as defined in Chapter I.

A = Equivalence ratio for the combustion of producer gas.

The JANAF program was partially utilized to execute the computations. It is noted that JANAF allows for the following options:

a) T, P specified

b) H, P specified

c) S, P specified

d) P specified, S same as last run

e) P specified, S and composition same as last run

f) T specified, S same as last run

g) H and P same as last run

h) T and P specified, composition same as last run

i) U and V specified

j) S and V specified

k) V specified, S same as last run

l) V specified, S and composition same as last run

m) U and V same as last run.

The same set of products as that used for the equilibrium calculations was allowed to form. The computation was executed along the following path.

a) Input of 100 kg of rice hulls at a given moisture content, M.

b) Determination of the amount of air needed at a given ϕ.

c) Determination of the atomic ratio of the rice hull-air mixture as $C_w H_x O_y N_z (SiO_2)_p$.

d) Computation of the equilibrium composition with JANAF option a) at p = 1 atm and T = 500°C - 1400°C.

e) Constant pressure expansion of the raw gas to a predetermined temperature T_f and computation of the enthalpy of the raw gas:

$$H_g = \sum_{i=1}^{n} H_{i,298} + \sum_{i=1}^{n} \int_{298}^{T_f} c_{pi} (T) \, dT \qquad (5-33)$$

T_f represents the temperature of the mixture of producer gas and air as fed into the burner. This step was done with JANAF option h).

f) Determination of the mass of raw gas generated over an input/output mass balance.

g) Determination of the stoichiometric amount of air needed to combust the raw gas.

h) Computation of the atomic fractions of the air-producer gas mixture for various A's. ($0.5 < A < 1.5$).

i) Computation of the total enthalpy H_R of the air-producer gas mixture at T_f.

j) Determination of the adiabatic flame temperature T_{ad} of the products when $H_R(T_f) = H_P(T_{ad})$. This last step was done with JANAF option b) with a two-dimensional iteration, guessing T_{ad} and then adjusting the product composition.

Flame Temperature as a Function of the Equivalence Ratio A

A typical raw gas composition:

$$CO = 17.1\%$$
$$CO_2 = 8.2\%$$
$$H_2 = 13.8\%$$
$$H_2O = 21.8\%$$
$$N_2 = 39.1\%$$

was assumed. For this composition the adiabatic flame temperature T_{ad} as a function of A is shown in Figure 5-24 at a mixture inlet temperature of 200°C. It is noted that in theory the peak temperature is reached at $A = 1$. In practice a higher A is needed to burn off most of the CO, hydrocarbons and solid carbon.

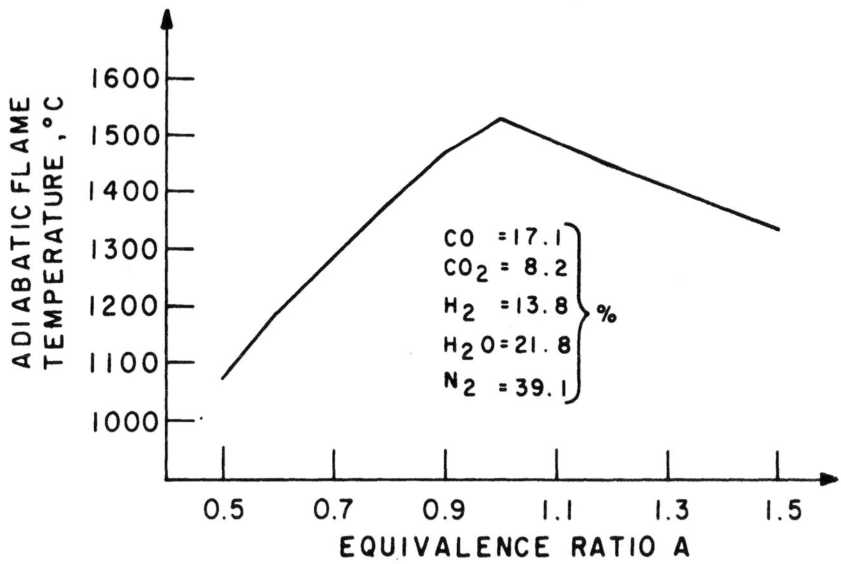

Figure 5-24. Adiabatic flame temperature as a function of A

Flame Temperature as a Function of the Products of Combustion

From a computational point of view, it is easiest to perform an energy balance if one assumes that the only products of combustion are CO_2, H_2O, O_2 and N_2. In reality, however, many radical and dissociation products are formed as listed in section 5.1.

The data for Figure 5-24 take into account the major dissociation reactions that occur in combustion processes. It is therefore of interest to determine whether the use of a restricted set of products such as CO_2, H_2O, O_2 and N_2 will significantly alter the results.

Computation of the adiabatic flame temperature under the latter assumption leads to $T_{ad} = 1542°C$ at $\lambda = 1$ as compared to $T_{ad} = 1530°C$ when dissociations are taken into account. The error from assuming a simple set of products is therefore smaller than 1% and is not significant.

Flame Temperature as a Function of the Mixture Inlet Temperature

In many applications it is desirable to feed the producer gas-air mixture at elevated temperatures into a furnace. There are several reasons for this practice:

a) The wide range of boiling points of the tarry products makes it necessary to avoid cooling down the gas too much before it is combusted.

b) Often the tar vapors in the raw gas are in a saturated state or slightly above the dew point. Mixing of producer gas with cold combustion air will therefore lead to the formation of tar droplets, which are more difficult to combust than tar vapor. These tar droplets may condense out along with the water formed during combustion and accumulate within the furnace.

c) In some applications it is desirable to reach furnace wall temperatures as high as possible (Stirling Engine, Carnot cycle).

The effect of the inlet temperature T_f of the gas-air mixture into the furnace is shown in Figure 5-25. It is noted that raising the inlet temperature of the mixture by 500°C will result in an increase in T_{ad} of only 325°C. This modest increase is not due to higher dissociation activity, as already shown. It is instead the result of the increase of $c_p(T)$ with temperature for all gaseous species.

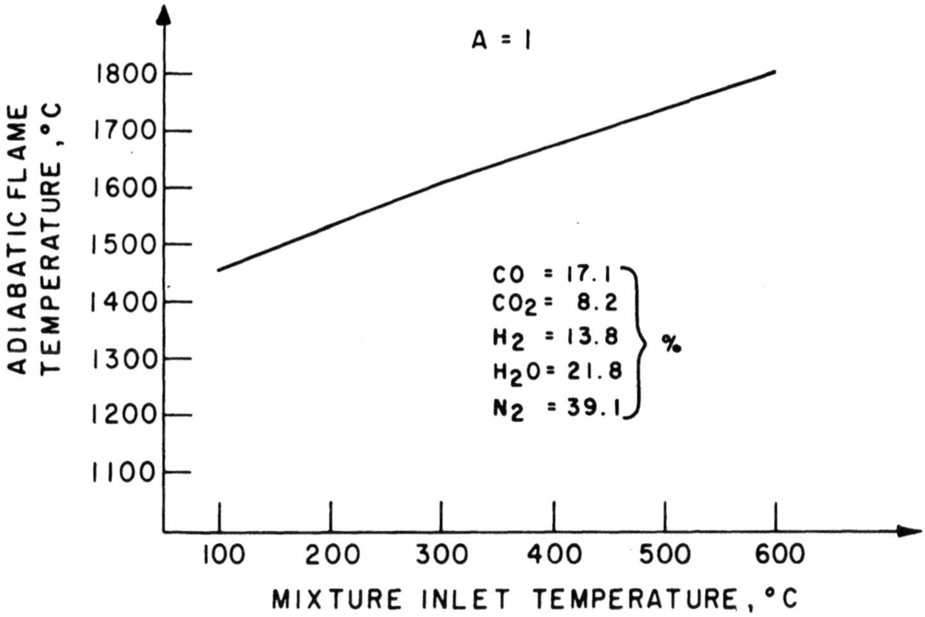

Figure 5-25. Adiabatic flame temperature T_{ad} as a function of mixture inlet temperature.

Flame Temperature as a Function of the Char Load in the Raw Gas

Direct firing of producer gas can not be assessed on the basis of a dry gas analysis, since the dust-laden, tar-laden gas is burned directly in a furnace. One obvious concern is whether large amounts of solid carbon in the raw gas stream will significantly alter the adiabatic flame temperature. The raw gas composition is the same as in the previous cases. Three cases: 0 grams, 6 grams and 20 grams of solid carbon per cubic meter of raw gas at an inlet temperature of 200°C were considered. The adiabatic flame temperatures for the three cases at A = 1, were:

$$T_{ad,0} = 1530°C$$

$$T_{ad,6} = 1568°C$$

$$T_{ad,20} = 1640°C$$

A significantly higher temperature may therefore be expected when high char loads occur in the raw gas.

Comparison of Theoretical Results and Experimental Data

One of the most extensive studies involving bulk phase temperatures of producer gas flames as well as wall and flue gas temperatures was conducted by Shoffstall, Waibel (1978). The results given in Table 5-5 were obtained by preheating the combustion air to 180°C and mixing it with producer gas at ambient temperatures.

Table 5-5. Adiabatic Flame Temperature and Measured Maximal Temperatures of Various Producer Gas Compositions

Gas	CO	H_2	CO_2	CH_4	N_2	H_2O	T_{ad}, °C	T_{max}, °C
Lurgi oxygen	18.5	40.2	29.4	9.4	0.6	1.9	1735	1541
Winkler oxygen	32.9	41.2	20.0	3.0	1.0	1.9	1830	1634
K-T-oxygen	52.1	34.5	9.2	0.5	1.0	1.9	1970	1782
Wellman-air	26.3	14.3	7.4	2.6	46.9	1.9	1620	1535
Winkler air	21.1	13.0	6.9	0.6	56.5	1.9	1415	1402

From the data one can conclude that the maximum temperatures obtained were very close for air blown gasifiers to the computed adiabatic flame temperatures. For example, the calculated T_{ad} for the Winkler Air case is 1415°C as compared to a measured maximum temperature of 1402°C. In addition, furnace wall temperatures of 1090°C and flue gas temperatures of 1285°C were reported in this case. It is noted that although producer gas heating values are only a fraction of the heating values of natural gas (15%-20%), adiabatic flame temperatures and measured temperatures in a furnace are only 200-300°C lower.

5. LIST OF SYMBOLS

A equivalence ratio for combustion of producer gas

$A_{\beta\sigma}$ total surface area of rice hulls within the control volume V (m^2)

$a_{\beta\sigma}$ specific surface area of rice hulls (m^2/m^3)

c_p specific heat of mixture at constant pressure ($J\ kg^{-1}\ K^{-1}$)

c_i molar concentration of species i

c_i surface concentration of species i (g-mole/m^2)

$c_{pi}(T)$ specific heat of species i at constant pressure ($J\ kg^{-1}\ K^{-1}$)

D_{zz} effective diffusion coefficient

H_g total enthalpy of producer gas mixture ($J\ kg^{-1}$)

$H_{i,298}$ heat of formation of species i at 298 K ($J\ kg^{-1}$)

H_i specific molar enthalpy of species i

H_p total enthalpy of the products of gasification or combustion ($J\ kg^{-1}$)

H_R total enthalpy of the products of gasification or combustion ($J\ kg^{-1}$)

J_i molar diffussive flux of specie i ($J\ m^{-2}\ s^{-1}$

$K_p(T)$ equilibrium constant at pressure p and temperature T

k surface reaction rate parameter

M moisture content of rice hulls (% weight, wet basis)

$\mathbf{n}_{\beta\sigma}$ orientation of the interphase area

P pressure of gas mixture ($N\ m^{-2}$

$\mathbf{q}$ heat flux vector due to conductive energy transport ($J\ m^{-2}\ s^{-1}$)

R universal gas constant (J/g-mole K)

R_i rate of formation or consumption of species i due to bulk phase reactions (g-mole/s)

T_f temperature of producer gas-air mixture (°C)

T_{ad}	adiabatic flame temperature (°C)
$\mathbf{t}$	stress tensor
$\mathbf{v}_i$	specie velocity vector
v	velocity vector of the bulk phase gas mixture ($m\ s^{-1}$)
$\mathbf{v}_i$	spatial deviation of specie velocity vector
V	control volume (m^3)
V_β	void space volume (m^3)
$\mathbf{w}$	velocity vector of the interphase area
β	coefficient of thermal expansion at constant pressure (K^{-1})
ρ	density of gas mixture ($kg\ m^{-3}$)
ε_β	void space fraction of fuel bed
$\langle f \rangle$	volume averaged parameter or physical constant f
Φ	energy source due to radiation ($J\ m^{-3}\ s^{-1}$)
ϕ	equivalence ratio for gasification of rice hulls
κ	linear adsorption isotherm

5. REFERENCES

1. Beagle, E. 1978. Rice husk conversion to energy. FAO Agricultural Services Bulletin #31. Rome.

2. Groeneveld, M.J. 1980. The co-current moving bed gasifier. PhD Thesis, Technische Hogeschool Twente, Netherlands.

3. Carbonell, R.G. and S. Whitaker. 1982. Heat and mass transport in porous media. Proceedings of NATO Advanced Study Institute of the Mechanics of Fluids in Porous Media, University of Delaware. July 18-27.

4. JANAF. 1971. Thermochemical Tables. 2nd Edition, U.S. Department of Commerce, National Bureau of Standards.

5. Probstein, R.F. and R.E. Hicks. 1982. Synthetic Fuels. McGraw-Hill. New York.

6. STANJAN Program. 1981. R. Reynolds Thermosciences Division, Department of Mechanical Engineering, Stanford University, Stanford, California 94305.

7. Shafizadeh, F. et al. 1976. Thermal Uses and Properties of Carbohydrate and Lignines. Academic Press. New York.

8. Shaffstall, D.R. and R.T. Waibel. 1978. Low-Btu gas combustion research. Institute of Gas Technology, Chicago, Illinois.

9. Van den Aarsen, F.G. et al. 1982. Performance of a rice husk fueled fluidized bed pilot plant gasifier. First International Producer Gas Conference, Colombo, Sri Lanka. November.

10. Whitaker, S. 1982. Lecture notes: Reactor design. Department of Chemical Engineering, University of California, Davis.

6. PHYSICAL PROPERTIES OF RICE HULLS

Densities

Introduction

As outlined previously, gasification is a complicated interaction of a gas-solid phase system. Based on a literature study and a survey of existing units, it is concluded that the physical properties of a fuel are usually underestimated while the influence of the heating values and ash content are overestimated with regard to the performance of a gas producer.

In the author's opinion, fuel properties should be ranked as follows with respect to their importance on the producer gas quality:

a) Physical shape of particle and fuel bed structure

b) Moisture content

c) Volatile matter

d) Ash content

e) Ash composition (natural and tramp ash)

f) Energy content.

The reasons for the above ranking will be outlined in subsequent chapters. This chapter will examine the fuel bed structure of a rice hull gas producer in detail.

Figure 6-1 shows a control volume V which may be viewed as the entire fuel column of a gas producer or only a very small section of it. An enlarged schematic sketch of a single rice hull and its various phases is given in Figure 6-2. Four phases are identified:

a) The β-phase having a volume V_β represents the void space within

Figure 6-1. Fuel bed structure of rice hulls.

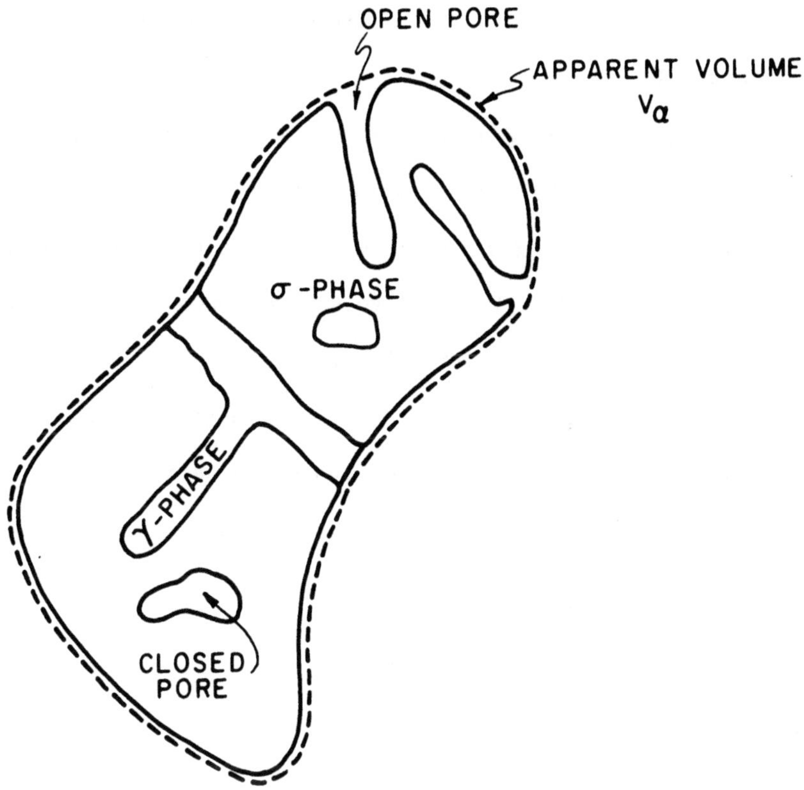

Figure 6-2. Apparent volume V_α of a single rice hull, and γ, σ phase.

the fuel bed if the rice hulls are treated as a nonporous material.

b) The α-phase having a volume V_α is the region the rice hulls occupy within the control volume V.

c) The γ-phase having a volume V_γ is the total pore volume of the rice hulls within V. The volume V_γ includes both the closed and open pores of the rice hulls.

d) The σ-phase represents the true solid matter of the rice hulls.

The various fractions of each phase are defined as:

$$\varepsilon_i = \frac{V_i}{V}$$

where $i = \alpha, \beta, \gamma, \sigma$.

It is noted that:

$$\varepsilon_\gamma + \varepsilon_\sigma + \varepsilon_\beta = 1$$

One can experimentally determine the following densities:

a) Bulk density: $\rho_b = \frac{m}{V}$

b) Apparent density: $\rho_a = \frac{m}{V_\alpha}$

c) True density: $\rho_t = \frac{m}{V_\sigma}$

The above densities were measured for rice hulls and pelletized rice hulls. The pellets were made by a ring-die pellet mill producing a 9.5 mm diameter pellet of length 20-22 mm. A series of over 130 measurements over a period of 1 year were taken. All data given are on a dry basis.

Bulk density ρ_b

The bulk density ρ_b. was determined in a plastic cylinder of known volume V = 6914 cm^3 and a height to diameter ratio of 3:1. This ratio corresponds roughly to the height - diameter ratios of small gas producers. The following average bulk densities for pellets ρ_{bp} and rice hulls ρ_{bh} were measured:

ρ_{bp} = 0.58 g cm^{-3}, range 0.55-0.60

ρ_{ph} = 0.10 g cm^{-3}, range 0.09-0.11

True density ρ_t

The true density, as well as the apparent density, were measured with a Beckman Air Comparison Pycnometer Model 930. True density measurements have been made by Tucker (1944). Beagle (1978) and Houston (1976) are assumed to have used Tucker's numbers in their publications. However, the fact that a single rice hull when immersed in water eventually sinks is indisputable proof that the reported true density ρ_t = 0.735 g cm^{-3} is in error. Attempts were made to repeat Tucker's tests with water and with toluene, which is a better wetting agent. The underlying principle of the test is to fill the pores with water and measure the volume change in the water bath. This method did not work, however, because too much air was trapped in the pores. Applying long-term vacuum to release the trapped air did not help. Thus, a more sophisticated method, air pycnometry, was employed.

In an air pycnometer, a weighed sample of rice hulls is exposed to a gauge pressure of 1 atm in a container of known volume. The air is allowed to diffuse into the open pores and the volume displacement by the true solid phase plus the closed pores is measured with a precision piston displacement within an accuracy of 0.05 cm^3 and better. The accuracy of the instrument is frequently tested with various precision steel balls of known volume. It is noted that the "real" true density of rice hulls can not be measured due to the following complications:

a) The fuel possesses closed pores inaccessible to air.

b) Micropores open to the air require a long time (sometimes many hours) to reach pressure equilibrium. Because the measurements are based on the air volume comparison at 1 atm gauge pressure, the errors introduced through prolonged pressurizing of the instrument can be significant and result in erratic data.

c) There is always the possibility of surface reactions, depending on the gas used for the measurement.

d) Any size reduction by milling destroys only an insignificant number of pores because the finest available sieves separate particles larger than 35 μm. This is a large fraction of the average thickness of a rice hull, which is 50 μm.

The measured true density, ρ_t, on a dry basis is therefore defined as:

$$\rho_t = x_1 \left(1 - \frac{x_2}{100}\right) \left(V - \frac{x_1 x_2}{100\ \rho_w}\right) \qquad (6-1)$$

x_1 = sample weight (wet basis, gram)

x_2 = moisture content in percent (wet basis, gram)

V = measured volume (cm^3)

ρ_w = density of water, 1 g cm^{-3}.

The air was allowed to diffuse into the pellets and rice hulls for at most 30 minutes and for at least 5 minutes before a measurement was taken. Eighty-three measurements of the sample volume V, with the Beckman Pycnometer using 6 different lots of pellets and hulls resulted in:

$\rho_t = 1.41$ g cm^{-3}, average dry basis; range 1.37 to 1.45 for rice hulls

$\rho_t = 1.61$ g cm^{-3}, average dry basis; range 1.58 to 1.63 for pellets.

The considerably higher true density for pellets is caused by compaction and by the destruction of a large number of closed and open pores during the pelletizing process.

For comparison, 14 tests using 3 different samples of rice hulls ground so that they passed through a 60 mesh screen (250 μ particles) resulted in:

$\rho_t = 1.60$ g cm^{-3}, average dry basis; range 1.59 to 1.61

A lower bound for the true density may be computed from an elemental ultimate analysis of rice hulls. The underlying assumption is the conservation of mass and complete combustion of the fuel to CO_2 and H_2O. Table 6-1 lists the mass fractions for rice hulls on a dry basis of C, O, H, N, Si, Fe, Mg, Ca, Zn, K, Na, S, and P. These constituents of rice hulls account for 99.9% of the mass of rice hulls. Based on the ultimate analysis and on the conservation of mass, the computed true density of dry rice hulls is $\rho_t = 1.56$ g cm^{-3}. Because the densities of H, O, N within the rice hulls are not known it was assumed that H and O are represented by water and we have therefore:

$$\rho_t = \sum_{i=1}^{n} x_i \rho_i$$

where x_i are the mass fractions from Table 6-1 and ρ_i the species densities. The small amount of nitrogen was neglected in the computation.

It should be pointed out that true densities reported on a wet basis are always lower and differ drastically due to the moisture in capillary openings, which will give a distorted value of ρ_t.

Table 6-1. Ultimate Analysis of Rice Hulls and True Densities of the Elements

Element	Mass Fraction % Dry Basis	True density, ρ_t* g cm^{-3}
C	41.44	1.8-2.26
O	37.32	-
Si	14.66	2.0-2.5
H	4.94	-
K	0.59	0.86
N	0.57	-
S	0.30	2.0-2.9
P	0.07	1.82-2.69
Ca	0.06	1.55
Na	0.035	0.97
Fe	0.006	7.86
Mg	0.003	1.74
Zn	0.006	7.14

*Data from Handbook of Chemistry and Physics, 62nd Edition, CRC Press.

Apparent Density ρ_a

The apparent volume, V_α, is defined as shown in Figure 6-2. It includes the solid matter as well as all open and closed pores. In order to measure the apparent volume of a sample of rice hulls it is necessary to close all pores that terminate at the surface. This is done by filling the pores with wax that is highly viscous at 125°C. Due to the tedious procedure only two samples of 1000 rice hulls each were tested. It is relatively easy to remove excess paraffin from a large object such as pellets. One can therefore compute the apparent density by:

$$\rho_a = \frac{m}{V_m} \tag{6-2}$$

where m = mass of unwaxed sample

V_m = volume measured of waxed sample.

However, equation (6-2) may not be used with small objects such as rice hulls. Electron microscope pictures showed that samples differed significantly with regard to the amount of visible excess paraffin on the surface and in the pores. If one assumes that no excess paraffin exists, the apparent density will be underestimated because a real sample is more accurately characterized by Figure 6-4. The ideal case is shown in Figure 6-3.

A more accurate determination of the apparent density may be obtained from ρ_t and the paraffin density ρ_p as follows:

The measured volume is: $V_m = Z_1 + Z_2 + Z_3$ (6-3)

according to Figure 6-5.

The apparent volume is: $V_\alpha = Z_1 + Z_2$

Thus, the true volume is $V_t = V_m - (Z_2 + Z_3)$ from which the apparent density on a dry basis is:

$$\rho_t = \frac{x_1(1 - \frac{x_2}{100})}{V_t} \tag{6-4}$$

Consequently, since $Z_2 + Z_3 = \dfrac{x_{2p} - x_1 (1 - \frac{x_2}{100})}{\rho_p}$ (6-5)

we get:

$$\rho_t = x_1 (1 - \frac{x_2}{100}) \left(V_m - \frac{x_{2p} - x_1 (1 - \frac{x_2}{100})}{\rho_p}\right)^{-1} \tag{6-6}$$

where V_m = measured volume of waxed rice hulls, (cm^3)

x_{2p} = sample weight, waxed rice hulls, (gram)

x_1 = sample weight, unwaxed wet basis, (gram)

x_2 = moisture content in percent.

ρ_p = density of paraffin = 0.888 g cm^{-3}.

It is noted that ρ_t is known with high accuracy. One can therefore compute a new V_m and x_{2p} that agrees with the measured ρ_t. The apparent density ρ_a is then given as:

$$\rho_a = \frac{m}{V_\alpha}$$

where $V_\alpha = V_m$. A series of 34 measurements resulted in:

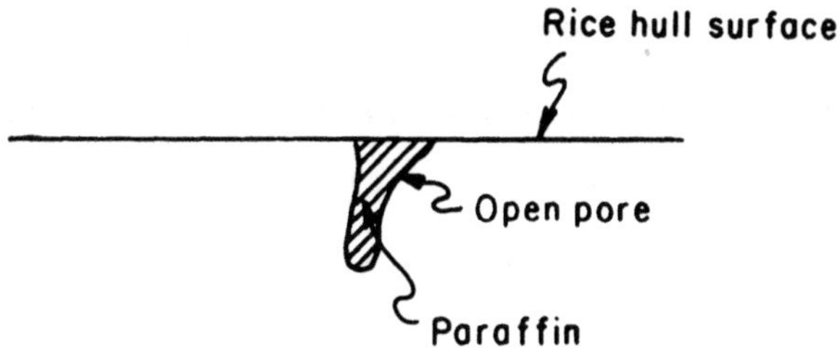

Figure 6-3. Ideal surface of waxed rice hulls.

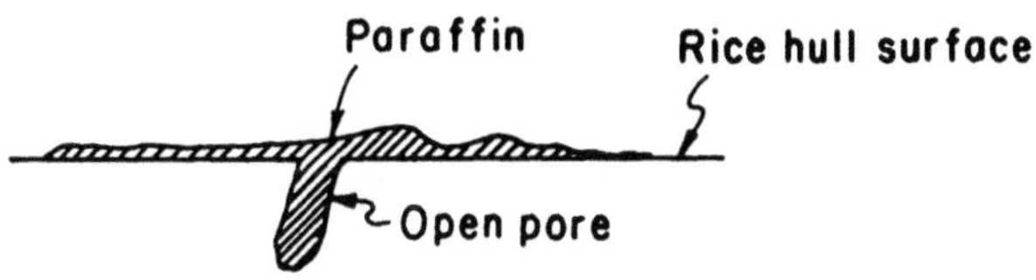

Figure 6-4. Actual surface of waxed rice hulls.

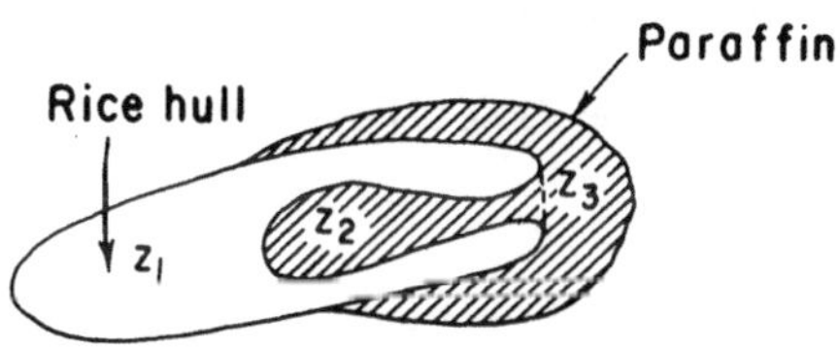

Figure 6-5. Geometry of waxed rice hull for computation of ρ_a.

ρ_a = 1.18 g cm^{-3}, dry basis, range 1.14-1.22 for pellets

ρ_a = 0.65 g cm^{-3}, dry basis, range 0.60-0.72 for rice hulls.

Phase fractions ε_α, ε_β, ε_γ, ε_σ

The phase fractions ε_i can be computed as follows:

$$V_\alpha = V - V_\beta \text{ and } \varepsilon_\alpha = 1 - \frac{V_\beta}{V} = 1 - \varepsilon_\beta \quad (6\text{-}7)$$

$$V_\beta = V - V_\alpha \text{ and } \varepsilon_\beta = 1 - \frac{V_\alpha}{V} = 1 - \frac{\rho_b}{\rho_a} \quad (6\text{-}8)$$

$$V_\gamma = V - V_\beta - V_\sigma \text{ and } \varepsilon_\gamma = 1 - \frac{V_\beta}{V} - \frac{V_\sigma}{V} = 1 - \varepsilon_\beta - \frac{\rho_b}{\rho_t} \quad (6\text{-}9)$$

$$V_\sigma = V - V_\gamma - V_\beta \text{ and } \varepsilon_\sigma = \frac{\rho_t}{\rho_b} \quad (6\text{-}10)$$

In practice, three porosities are important:

a) The total void space fraction $\varepsilon_t = \varepsilon_\beta + \varepsilon_\gamma$ (6-11)

b) The porosity of the fuel $\varepsilon_f = 1 - \frac{\rho_a}{\rho_t}$ (6-12)

c) The void space fraction between the rice hulls

$$\varepsilon_\beta = 1 - \frac{\rho_b}{\rho_a} \quad (6\text{-}13)$$

These porosities are given in Table 6-2 for rice hulls and rice hull pellets.

Table 6-2. Gas-Solid Phase Distribution in a Gas Producer for Loose Hulls and Rice Hull Pellets

	Pellets	Hulls
True Density, ρ_t, g cm^{-3}	1.60	1.42
Apparent Density, ρ_a, g cm^{-3}	1.18	0.65
Bulk Density, ρ_b, g cm^{-3}	0.58	0.10
ε_f	26%	54%
ε_β	51%	85%
ε_t	64%	93%

Surface Area of Loose Rice Hulls and Rice Hull Pellets

Introduction

Since most gasification reactions are heterogeneous, the specific surface area $A_{\beta\sigma}$ (m^2/m^3) plays an important role in the process because of its influence on the reactivity of the fuel bed. It is noted that the apparent surface $A_{\beta\sigma}$ refers to the surface area one would find if the rice hulls were nonporous, as indicated by V_α in Figure 6-2. This surface area is measurable with an image analyzer and electron microscope. The real solid-gas interface in a porous media is usually determined by the BET method (Brunauer-Emmett-Taylor). This surface area includes all internal surfaces resulting from the porous structure of rice hulls. It can be startlingly high, on the order of 10^6 m^2 kg^{-1} for activated charcoal from coconut shell, wood, and bituminous coal (Mantell, 1968).

Determination of the BET surface is based on vapor adsorption and is beset with difficulties. The theory is essentially an extension of the Langmuir theory, which is based on two very tenuous postulates:

a) The adsorbed vapor covers the entire surface with a uniform monolayer.

b) The amount of adsorbed vapor can be detected from the adsorption isotherm.

The assumption of a monolayer can hardly be justified and even extending the theory to multilayer adsorption (BET surface) leaves some doubt as to the accuracy of the method.

Any useful determination of the solid-gas interphase area at various stages in the gasification process of rice hulls must certainly be preceded by the development of an appropriate method of obtaining reliable samples and of evaluating the maximum error for surface measurements. There does not appear to be a convincing solution to these problems at present.

Determination of $A_{\beta\sigma}$ for rice hulls

The apparent surface area $A_{\beta\sigma}$ of a single rice hull has been experimentally determined under the following assumptions and procedures:

a) Electron microscope pictures of cross sections were used to fit a curve that will describe the regular appearance of the geometrical bumps on the outer surface of the rice hulls. The shape of the bumps, as shown in the electron microscope photographs (Figure 7-4 is an example) was measured at 6 points. The shape was approximated by a sixth-degree polynomial,

$$P(x) = \sum_{i=0}^{6} a_i x^i \qquad (6\text{-}14)$$

The assumed cross section is shown in Figure 6-7 with the mean measured dimensions.

b) The bumps were assumed to be rotationally symmetric around their vertical Y axes. The surface area of each bump was computed by

$$S = \int_0^{36} 2\pi x(1 + [P'(x)]^2)\, dx \qquad (6\text{-}15)$$

c) A composite picture, consisting of 6 sections at a magnification of 25 was then used to establish the total enclosed plane area of three single rice hulls. This circumference was then traced by a computer-guided image analyzer that automatically computed the enclosed area.

d) The number of bumps present on the outer surface was established by counting them and checked by superimposing a 72 μ grid over the composite picture.

It is noted that a "single" rice hull looks as shown in Figure 6-6. The rice hull consists of two parts, each of which can be naturally separated into two additional parts. The average computed apparent surface area was $A_{\beta\sigma} = 3855\ m^2/m^3$ and is based on the computed cross section shown in Figure 6-7 and ρ_{bh} previously given. It is noted that the roughness of the outer surface (bumps) increases the total outer surface area by about 12%.

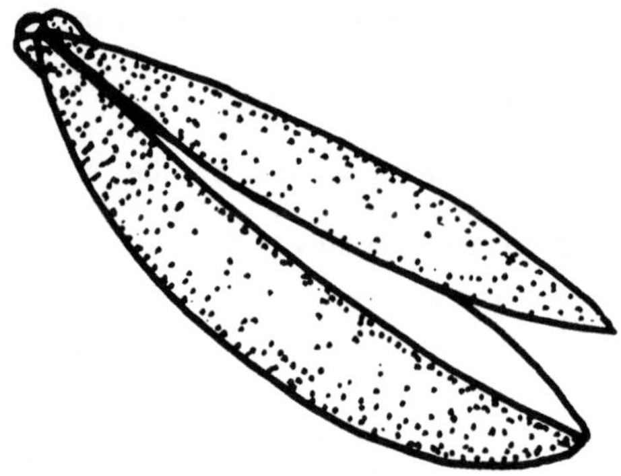

Figure 6-6. Schematic of a single rice hull.

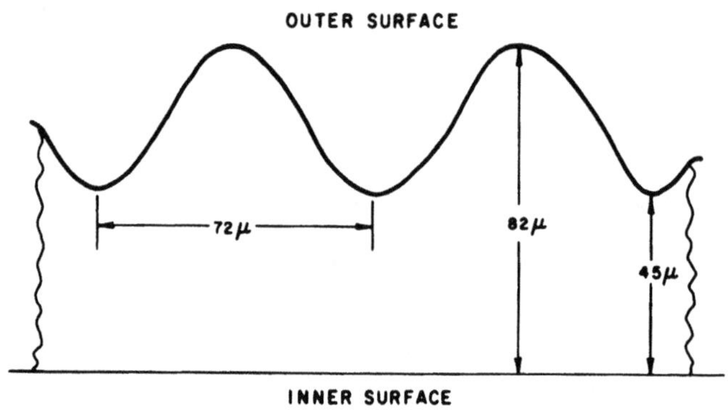

Figure 6-7. Modelled cross section of rice hull.

Apparent Surface $A_{\beta\sigma}$ of Pellets

The surface area $A_{\beta\sigma}$ of pellets has been experimentally determined using the following assumptions and procedures:

a) The pellets were made with a ring die a pelletizer that is normally used to manufacture alfalfa hay pellets for livestock.

b) The particular die ring used generates pellets with a diameter of 9.6 mm and an average length of 22 mm. However, during subsequent handling a considerable number of pellets were broken into pieces of various lengths. It was found that approximately 30% of the pellets remained unbroken and that 70% had been broken.

c) The additional surface area generated by breaking a single pellet into two pieces has been taken into account. The average apparent surface area was $A_{\beta\sigma} = 302\ m^2/m^3$.

Weight of a Single Rice Hull

In order to obtain a complete description of the geometry of a rice hull bed in a gas producer, determination of the average weight of a single rice hull is necessary. The measurements were based on a selection of 50 single rice hulls that had both hull parts, as shown in Figure 6-6. Weight measurements were performed with a Kahn-Electro Balance with an accuracy of $\pm\ 10^{-6}$ gram. The range for these rice hulls was $m_s = 2.944$ to 3.564 mg (dry basis) with an average m_s of 3.087 mg.

One of the difficulties in measuring the weight of a highly porous small object such as a rice hull is its moisture content. The measuring chamber of the electro-balance is lighted; thus, there is a heat source which evaporates small amounts of equilibrium moisture within seconds. The average decline in the rice hull weight during the measurement process was 30 μg.

Accuracy of the Results

Because a wide range of measurement techniques (some of them novel) have been used to characterize a rice hull fuel bed, it is appropriate to perform an error analysis for the data reported in Chapter 6.

All data has been reported on a dry basis to allow comparison of various samples and to eliminate the "moisture content" variable which can considerably distort experimental data.

The error analysis is based on the following:

a) Random errors are not included in the analysis.

b) Instrument errors and data transformation errors are based on manufacturer's data and on the type of instrument.

c) An upper bound for the maximal error has been computed in the following manner.

For instance, the dry bulk density, ρ_b, is given by:

$$\rho_b(x_1, x_2, x_3) = \frac{x_2 \left(1 - \frac{x_1}{100}\right)}{x_3} \tag{6-16}$$

where x_1 = sample moisture content in % wet basis

x_2 = mass of wet sample, gram

x_3 = volume of sample container, cm^3.

Furthermore,

$$\Delta\rho_b\,(x_1, x_2, x_3) = \sum \frac{\partial\rho_b}{\partial x_i}\Delta x_i \qquad (6\text{-}17)$$

and consequently,

$$\frac{\Delta\,\rho_b}{\rho_b} \leq \frac{\sum_{i=1}^{3} \frac{\partial\rho_b}{\partial x_i}\Delta x_i}{\rho_b} \qquad (6\text{-}18)$$

where $\frac{\Delta\,\rho_b}{\rho_b}$ is the largest possible error.

In this case,

$$\frac{\Delta\rho_b}{\rho_b} \leq \frac{\Delta x_1}{100-x_1} + \frac{\Delta x_2}{x_2} + \frac{\Delta x_3}{x_3} \qquad (6\text{-}19)$$

Assumptions:

Δx_1 = 0.5% (error in the moisture content measurement)

Δx_2 = 5 gram (error in wet sample)

Furthermore, assumed measurement errors of $\pm$ 1 mm for the diameter and height of the sample container with a total volume x_3 = 6914.26 cm^3 results in

$$\frac{\Delta x_3}{x_3} = 0.016$$

Finally,

$$\frac{\Delta\rho_b}{\rho_b} \leq 0.0055 + 0.0021 + 0.016 = 0.024 \widehat{=} 2.4\%$$

Based on this technique maximum errors in the data have been established and are listed in Table 6-3.

Table 6-3. Maximum Error Data for Rice Hulls and Rice Hull Pellets

Property	Maximum Error %	
	Pellets	Hulls
Bulk Density ρ_b	2.4	2.8
Apparent Density ρ_a	4.2	5.6
True Density ρ_t	1.3	3.4
Porosity ε_f	15.5	8.3
ε_β	6.4	1.4
ε_t	2.1	0.5
Weight of Single Rice Hull or Pellet	0.9	1.1
Surface Area	*	3.8

*insufficient data

Although electron microscope evaluations and image analyses have an error of less than 3%, only a few samples (3 in our case) can be evaluated because of the tedious procedures involved. This can hardly be viewed as a representative sample from a population of 33,000/liter. Some improvement in the accuracy may be achieved by correlating the surface area to the weight of a single rice hull based on the model given in Figure 6-7. From a measurement standpoint it is much easier to obtain a reliable weight distribution from a large sample population of rice hulls.

Caking and Slagging Behavior of Rice Hulls

Introduction

The present interest in utilizing a wide variety of biomass fuels for gasification has resulted in an increased awareness of the phase changes of solid fuels. However, inadequately presented results and inappropriate descriptions of the slagging process have resulted in some confusion regarding the process. It is therefore essential to clarify the differences between caking and slagging (including the causes of each) before the experimental results are presented.

Any solid material will change from a solid to a liquid or gas when heated up to high enough temperatures. Coal and biomass fuels are no exception and will melt down at high temperatures. For instance, one can envisage what would happen if a sample of rice hulls was placed in an inert atmosphere and heated. It is noted that rice hulls consist of C, H, O, N, SiO_2, K_2O, N_2O and some other trace minerals most likely in their oxidized form. The first elements that will "melt off" and change from the solid phase to the liquid and gas phases are H, O, N and some C in form of compounds. This process is called devolatilization or pyrolysis and will be discussed in Chapter 8 in detail. Pyrolysis results in the formation of free carbon radicals which recombine with the freed atomic hydrogen to form a liquid (tar, oils and water) that is easily evaporated at higher temperatures. In addition, a gas phase consisting mainly of CO and CO_2 is formed directly. At higher temperatures some of the low temperature mineral oxides such as K_2O and Na_2O will reach a transition state. In this state the minerals soften before becoming completely liquified. It is noted that despite the high temperatures, no carbon will be burned off since the atmosphere is inert. The final result is therefore a molten mass of minerals which entomb solid carbon and high temperature mineral oxides such as SiO_2, MgO and CaO. The melting points of selected minerals and mineral oxides as well as carbon are given in Table 6-4.

Table 6-4. Melting Point of Selected Oxidized Minerals

Mineral Oxide		Melting Point °C*
CaO		2614
MgO		2852
K_2O		350
Na_2O		1275
Fe_2O_3		1560
P_2O_5		580
SiO_2:	Quartz (amorphous)	1460-1610
	Trydimite	1703
	Cristobalite	1723
C:	amorph	3652

*Data taken from the Handbook of Chemistry and Physics, 62nd Edition, CRC Press, 1982.

It is noted that the softening and melting points of biomass must be determined experimentally because they are strongly dependent on the mass fractions of K_2O and Na_2O.

One could also envision placing a sample of rice hulls into a highly oxidizing atmosphere and adding heat. In this case all the carbon in the fuel will be quickly burned off. The residues from this process (minerals) will melt down if the temperature reaches the softening and melting temperatures of the particular sample.

A third phenomena is closely associated with the low temperature pyrolysis process. It has already been mentioned that unstable free radicals are formed in the 250-450°C range. Those which react with hydrogen from the fuel form tar, whereas the rest tend to form bonds with one another. In the case of rice hulls this latter process leads to formation of large fluffy conglomerates which are easily breakable yet which nevertheless obstruct gravity flow. This phenomena is called caking of the fuels. It is widely observed in the gasification and combustion of certain bituminous coals. Caking of coal was prevented by stirring the bed or by preoxidizing the coal, which results in a product known as coke. The notion of caking has so far not found a place in the biomass gasification literature.

In summation: Three forms of biomass conglomerates are sometimes found within the gas producer:

a) Conglomerates that have been generated in an inert atmosphere within the gasifier. These rock-like structures are made up of melted minerals and entrained carbon (slagging).

b) Conglomerates that have been generated in a highly oxidizing atmosphere within the gasifier, resulting in the meltdown of minerals (slagging).

c) Conglomerates that have been formed by free radicals which react with one another at low temperatures (caking).

Depending on the type of biomass and the conditions under which it is gasified, all three processes may occur. Also, some interaction does occur between these processes.

Slagging of Rice Hull Ash

According to some authors, rice hull ash is highly prone to slagging (Jenkins 1981, Goss 1980, Groeneveld 1982) due to its high silica content (16-22%) and alleged low mineral melting point (800-1000°C).

Based on this information, the author has tried unsuccessfully to build a rice hull gas producer which induces slagging of rice hulls. In the course of these attempts several tests were conducted that showed that rice hull ash has no tendency to slag below 1400°C if localized combustion with excess oxygen does not occur. Osman (1982) has measured the softening and melting point temperatures of various biomass ashes. His results are listed in Table 6-5. The experiments were conducted in a nitrogen atmosphere. The biomass ash was made into small cones and the softening temperature was defined as the temperature at which the peak of the cone toppled over and began to deform. It is noted that the softening temperature of the biomass fuel will be the lowest temperature at which the possibility of fuel flow obstruction exists.

From the table it is seen that the rice hull ash has the highest softening and melting point of all tested biomass fuel residues. It is noted that when present in large amounts low melting point temperature mineral oxides such as Na_2O and K_2O will significantly lower the melting point of the ashes. This is shown in the phase diagrams taken from Toropov (1972) for the Na_2O - SiO_2 and K_2O - SiO_2 system (Figures 6-8 and 6-9). It is noted that SiO_2 is found in three forms in rice hull ash: quartz, cristobalite and trydimite. Jones (1953) has already pointed out that rice hull ash contains a negligible proportion of quartz and a high proportion of cristobalite and trydimite. This was later confirmed by Bartha (1974). Jones (1953) also reported that rice hull ash, when mixed with 3-15% CaO (lime), showed no tendency to soften nor was there a significant change in the volume of the ash at temperatures of 1667°C.

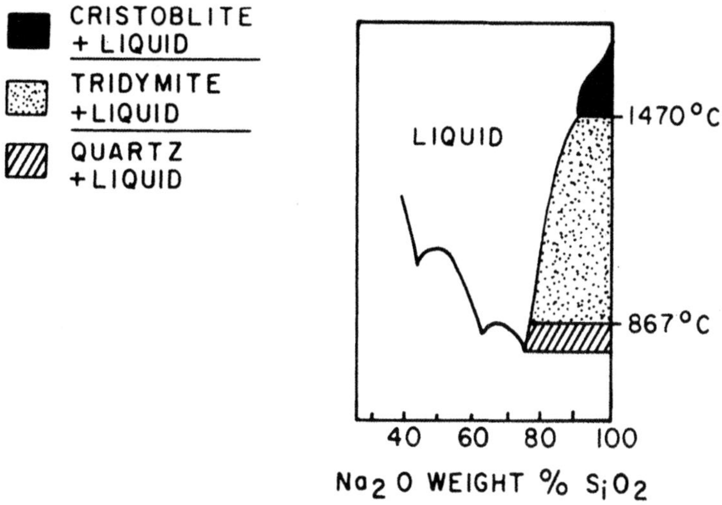

Figure 6-8. Na_2O - SiO_2 system (Levin, 1964).

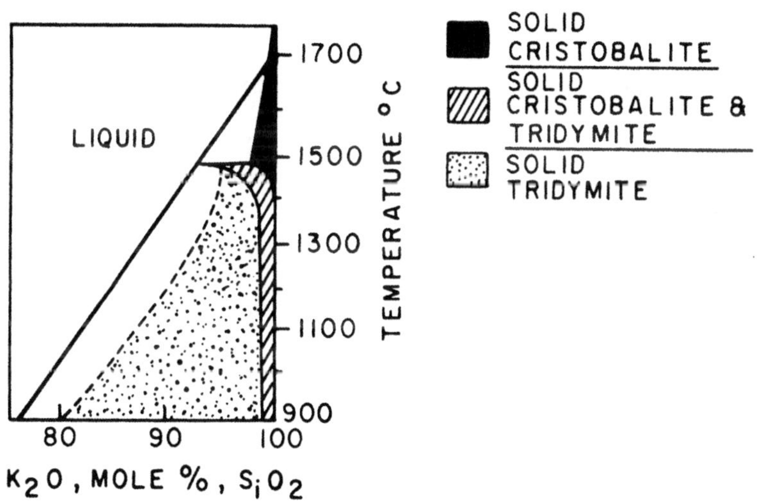

Figure 6-9. K_2O - SiO_2 system (Levin, 1964).

Table 6-5. Softening and Melting Temperatures of Biomass Ashes*

Biomass Ash	Softening °C	Melting °C
Bean straw	900	1150
Safflower	770	1430
Rice hulls	1439	>1650
Cotton gin trash	1010	1380
Barley straw	925	1100
Corn stalks	820	1091
Rice straw	823	1190
Wood chips	1050	1190
Corn fodder	1010	1180
Almond shell	790	1440
Corn cobs	900	1020
Tree prunings	770	1550
Walnut shell	820	1225
Olive pits	850	1480
Almond shell	860	1350
Good rice straw	1060	1250
Partially decomposed rice straw	850	1280

*sample values are highly dependent on ash composition (fluxing effect).

Because low melting point minerals affect softening point temperatures, it is of interest to look at the available data on rice hull ash composition collected by Beagle (1978), Houston (1976) and Kaupp, Goss (1982) shown in Table 6-6. This data represent values determined as long ago as 1910 and on four continents (North America, Europe, Asia, Australia). Comparing the listed values for K_2O and Na_2O with the phase diagrams in Figures 6-8 and 6-9 one can conclude that local variations

Table 6-6. Rice Hull Ash Composition

Specie	% Weight Fraction
SiO_2	90-97
CaO	0.2-1.5
MgO	0.1-2
K_2O	0.6-1.6
Na_2O	Traces-1.75
P_2O_5	0.3
SO_3	0.1-1.13
Cl	0.15-0.4
Fe_2O_3	Traces-0.4
MnO_2	Traces

in the ash composition will not result in a softening temperature below 1400°C. A review of the experimental data and operational reports for the large (60-100 hp) Italian rice hull gas producer (see Section 10-1) did not reveal a single reference to slagging problems due to the rotating grate and a low superficial velocity. Considering the available evidence, it is concluded that rice hulls will not slag under normal operation of a gas producer (temperatures during normal operation will always be under 1400°C). Nevertheless, severe slagging problems have been reported and there is a real possibility that slagging will occur in a rice hull gas producer as will be outlined later. However, the false conclusions put forth concerning the slagging potential of rice hulls result from a misinterpretation of the true cause of slagging, which has to do with the rice hull fuel bed structure and not directly with the high ash content of rice hulls or its silica content.

Cause of Slagging

In Chapter 6 the fuel bed structure of rice hulls was examined in detail. It was concluded that due to the high void space fraction ($\varepsilon_\beta \approx 0.85$) and low bulk density (100 kg m^{-3}), the homogeneous fuel bed structure is easily disturbed. From coal gasification it is known that channel formation within the coal bed, which can be caused by caking or by a large proportion of fine particles, is very detrimental to the gasification process. Channel formation usually leads to an erratic gas composition and also leads to the formation of hot spots. Because the size distribution of rice hulls is within a very narrow range (see Figure 6-21, Section 6.7) channel formation due to the presence of fine particles will not occur. However, caves and channels are readily formed in a rice hull fuel bed when settling creates shear zones (cracks) and the local superficial gas velocity exceeds the average value. Experiments to determine the cause of slagging have been performed in an updraft gas producer with a 30 cm firebox diameter and with a fixed flat grate (Figure 6-10). It is noted that gasification can not be sustained in such a gas producer because there is no continuous ash removal. The fire zone will therefore move upwards from the ignition point at the grate, consuming layer after layer of unburned rice hulls until the entire rice hull column has been charred. The combustion air flow rate was set at 50 Nm^3 h^{-1}, which represents a superficial velocity of at least 0.19 m s^{-1} within the column. The actual superficial velocity within the column is higher since the gas output is greater than the original air input by the amount of solid fuel converted to gaseous products. When the fire zone reached the gas exit port, the combustion

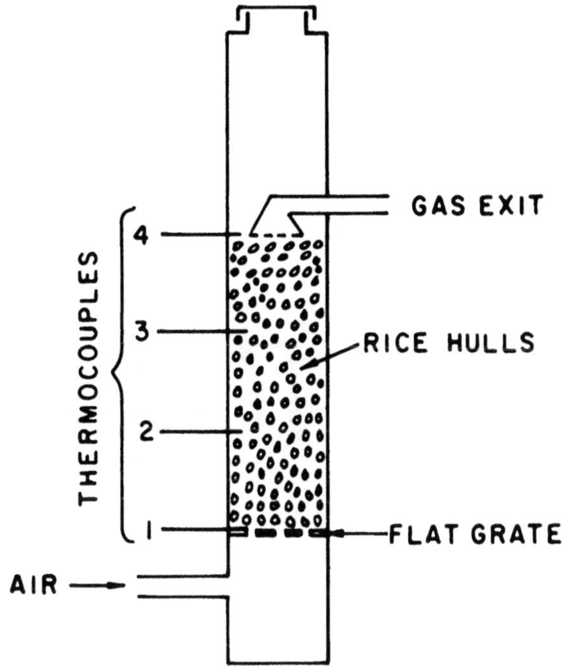

Figure 6-10. Experimental gas producer for testing the slagging behavior of rice hulls.

air was shut off and the fuel bed was examined layer by layer. A schematic sketch of the fuel bed structure is shown in Figure 6-11. The fuel bed developed characteristic caves that were glazed with a layer of molten snow-white rice hull ash. A sample of this form of slagging (as described in the introduction) is shown in Figure 6-12. The cause of this slag formation is clearly localized complete combustion of the rice hulls due to high localized airflow. The formation of a cave or channel allows air to reach the carbon surface at an equivalence ratio of $\phi > 1$. Because of the poor conductivity of the fuel bed and radiation within the caves, localized temperatures remain high, which results in melting of the rice hull ash. The installed thermocouples showed only normal gasification temperature of 900-1200°C. Thus, one might be tempted to

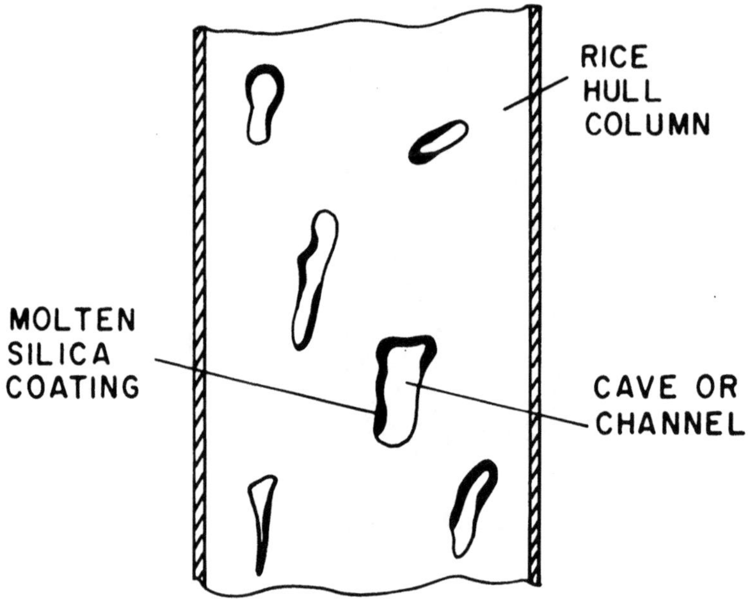

Figure 6-11. Schematic of slag formation in a rice hull fuel bed (caves not to scale).

wrongly conclude that slagging took place at low temperatures since the thermocouples were not placed where the slagging was actually occurring.

The same experiment was repeated with rice hull pellets. Although rice hull pellets do not easily allow channel formation, they do tend to form hot spots, which in isolated cases leads to melting of the ash which in turn leads to bonding between pellets as shown in Figure 6-13. It has been concluded that slag formation is caused by high superficial air velocities in the fuel bed which lead to localized meltdown (glazing) of the rice hull ash. In order to determine the approximate superficial velocity at which disturbances in the fuel bed can be expected, experiments in a cold rice hull and rice hull char fuel bed were conducted. The results are given in Section 6.6.

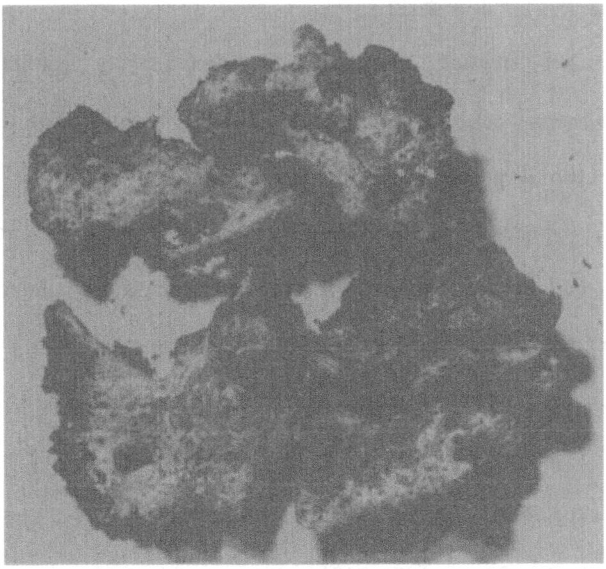

Figure 6-12. Molten silica and slag formation.

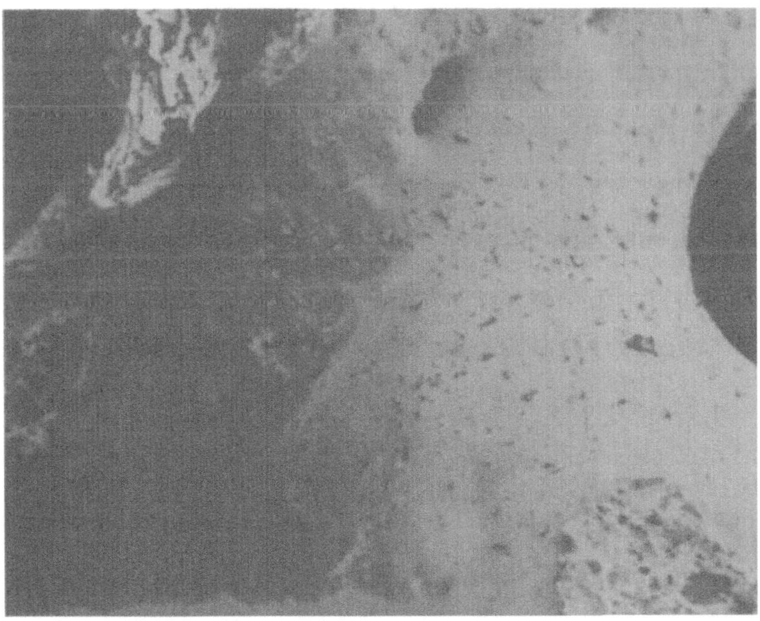

Figure 6-13. Localized slagging, connecting pellets of rice hulls.

Caking of Rice Hulls and Pellets

As previously argued, caking of biomass is strictly a low temperature phenomena caused by pyrolysis or thermal destruction of the fuel. It is also a phenomena that is closely related to the size of the gas producer, because caking can go unnoticed or can be prevented (through the use of mechanical stirrers) in larger units. The caking behavior of rice hulls is most easily demonstrated in an open core gas producer which operates without air nozzles as shown in Figure 10-27, Section 10-4. To induce caking, the gas producer is first fueled with a good gasification fuel, such as corncobs, until a sufficiently deep hot carbon layer develops. The gasification process is continued by the addition of rice hulls through the open top of the gas producer. A characteristic slight shrinkage of the rice hulls will take place throughout the entire rice hull column. In addition, a very distinct gap between the gas producer wall and the rice hull bed will develop. This effect is shown in Figures 6-14 and 6-15 which shows a top view of the rice hull fuel bed during thermal decomposition.

In order to make caking more visible, pellets can be used. A typical bond is shown in Figure 6-16 (black line). This figure should be compared to Figure 6-13 which shows a bond caused by local over-oxidation and meltdown of the ash.

Summary

Rice hull ash does have a very high softening temperature (> 1400°C), but nevertheless tends to slag if the fuel bed structure is disturbed by high superficial gas velocities. Rice hulls do cake during the pyrolysis process which may obstruct gravity flow. This property along with the insignificant size reduction of rice hulls during gasification are some principal disadvantages in rice hull gasification.

Figure 6-14. Initial stage of rice hull gasification.

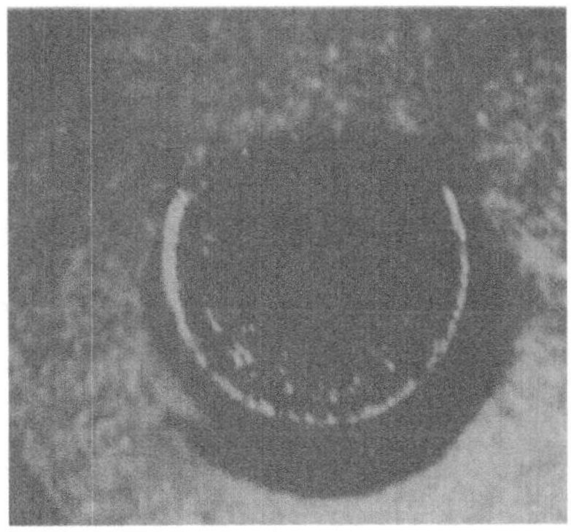

Figure 6-15. Final stage of rice hull gasification.

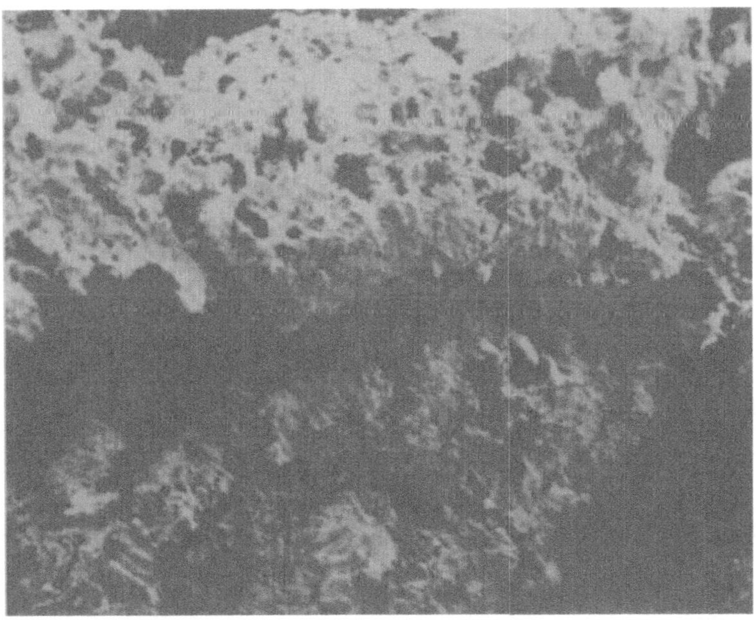

Figure 6-16. Caking of rice hull pellet at low temperatures (300 - 500°C).

Pressure Drop in a Rice Hull Fuel Bed and Superficial Velocities

It was outlined in the previous section that disturbance of the fuel bed will lead to slag formation and to erratic gas compositions. It is therefore of interest to determine the minimum gas velocity which will cause disturbances in a rice hull fuel bed and rice hull char bed. Also of importance is the pressure drop across such a system, because this pressure drop must be overcome by the intake manifold vacuum of a given engine, if the gas is to be used in the engine without special pumping. It is noted that the Italian rice hull gas producers were 1.5 m in diameter to power 60-100 hp engines (see Section 10.1). It is assumed that 3 Nm^3 of producer gas are necessary to generate one brake-horse power hour and that the gas will be at approximately 200°C when leaving the upper layers of the rice hull bed. Under these assumptions, the superficial gas velocity will be 7.5 cm s^{-1}. If the computation is based on the air entering the lower charred layers in an updraft gas producer, then the resulting superficial gas velocity will be 10 cm s^{-1}. The latter computation is based on the following: (1) 2 kg of rice hulls are needed in order to generate one brake-horse power hour; (2) the equivalence ratio is $\phi = 0.4$; (3) air enters the fuel bed at a temperature of 600°C. It is concluded that in the past, updraft rice hull gas producers operated at very low superficial velocities, on the order of 0.1 m s^{-1} (throughout the fuel bed). Downdraft gas producers operated on wood blocks have superficial gas velocities of 0.5-3 m s^{-1} in the fuel bed, which are considerably higher (Nordstroem 1957, Kaupp and Goss 1982).

Experimental Set Up and Results

The pressure drop across a rice hull and rice hull char fuel bed was measured in a vertical tube as shown in Figure 6-17. The rice hulls were supported by a fine (3 mm) wire mesh. The volumetric flow rate of the incoming air was measured with a calibrated sharp edge orifice plate (4% accuracy). The pressure drop across the bed was measured with an inclined manometer ($\pm$ 2.5 mm H_2O reading accuracy). The wire mesh induced a pressure drop of less than 2.5 mm H_2O at all tested flow rates. At the maximum flow rate the corresponding Reynolds number was 2200, which indicates the presence of a laminar flow field.

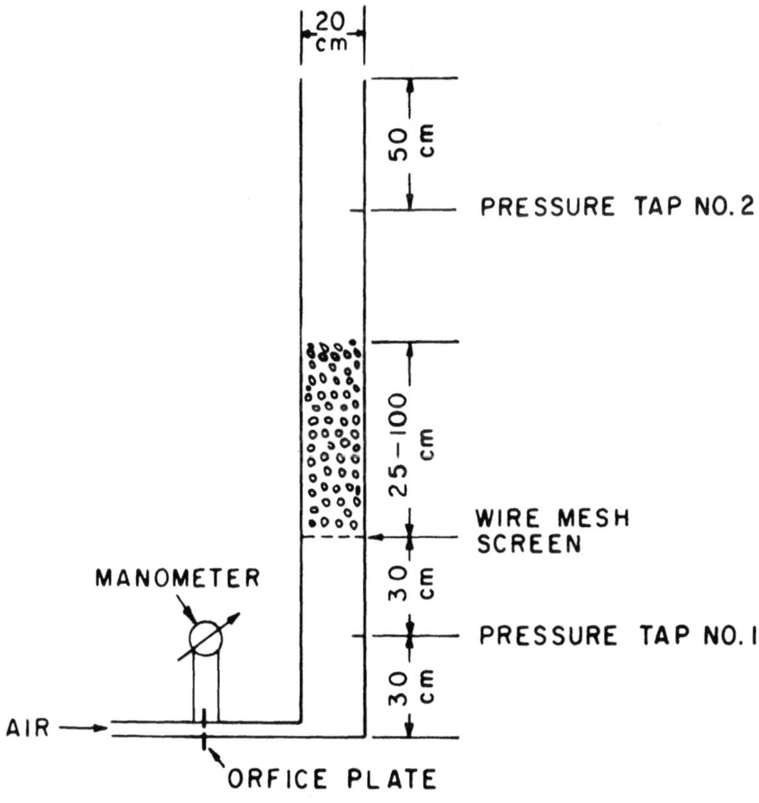

Figure 6-17. Experimental set up for pressure drop testing.

Tests were conducted using 25 cm, 50 cm, 75 cm and 100 cm deep unburned rice hull and rice hull char beds. The objective of the experiment was three fold:

a) To determine the pressure drop across beds of various lengths;

b) To determine the superficial velocity which results in channel formation, using visual inspection;

c) To compare the results for unburned rice hulls and rice hull char.

The results of the experiments are given in Figure 6-18.

Discussion of the Results

a) Channel Formation

For the purposes of this experiment, one must distinguish between the fluidizing velocity, v_s, and the velocity at which channel formation takes place. The velocity at which single unburned rice hulls or charred rice hulls were lifted up from the bed is defined as the fluidizing velocity. Channel formation does not necessarily start at this velocity. Based on the observations, the fluidization of single rice hulls (not broken rice hulls) was always accompanied by channel formation. However, in the case of rice hull char, channel formation started at a lower superficial gas velocity than the fluidization of individual charred rice hulls. The following conclusions were drawn:

i) Severe visible channel formation occurred at superficial air velocities of 8.5-9 cm s^{-1} for charred rice hulls. Channel formation was accompanied by a reduction

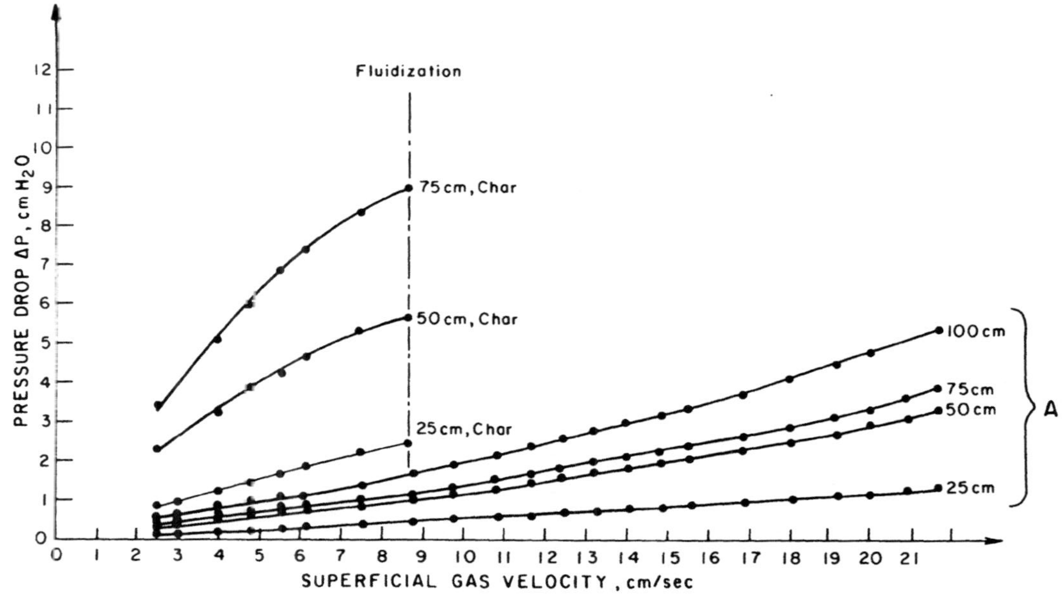

Figure 6-18. Pressure drop through an unburned rice hull bed (A) and rice hull char bed as a function of the superficial gas velocity v_s, for various bed lengths.

in the pressure drop despite an increased flow rate.

ii) The convex shape of the curves for rice hull char indicates that slight disturbances in the fuel bed must have taken place at lower superficial gas velocities.

iii) The unburned rice hull bed was considerably more stable than the char bed. Rice hull fluidization and channel formation began at a v_s of 20-23 cm s^{-1}.

iv) Initial fluidization and channel formation never occurred at the center or at the wall of the tube. Instead these phenomena always began at a short distance away from the wall. This observation may suggest a horizontal velocity profile within the bed as shown in Figure 6-19.

Based on these results it is understandable why classic Italian

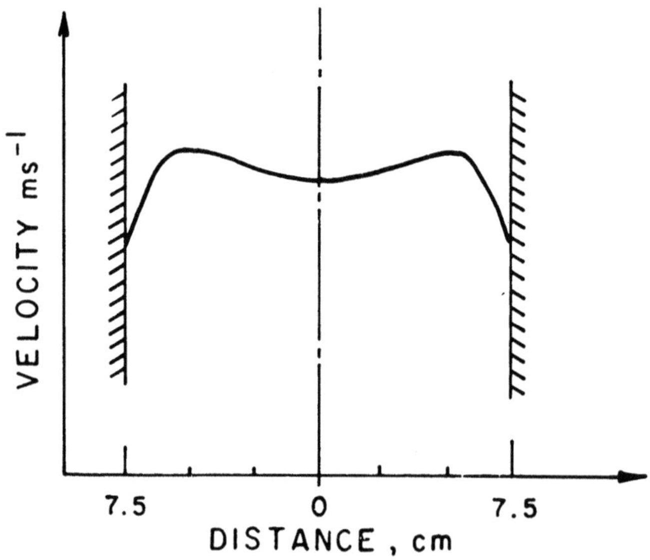

Figure 6-19. Probable horizontal gas velocity profile within a rice hull bed.

rice hull updraft gas producers were operated at such low superficial gas velocities and why they were made so large as compared to today's compact downdraft gas producers.

b) Pressure drop

The pressure drop (reported as a function of the superficial velocity v_s) was measured up to the fluidizing velocity. The results are shown in Figure 6-18. The pressure drop Δp can be viewed as the following function:

$$\Delta_\rho = \Delta p\ (\varepsilon_\beta, \rho, V_s, L, \gamma, D_p)$$

where

ε_β = void space fraction

ρ = gas density, kg m^{-3}

v_s = superficial gas velocity, m s^{-1}

L = length of rice hull column, m

γ = surface roughness

D_p = particle diameter, m.

Intutitively we should have $\frac{\partial \Delta p}{\partial L}$ = constant, in the absence of entrance and exit disturbances. The rice hull char bed follows this law closely while the unburned rice hull bed data indicates that large entrance and exit disturbances must be present. Although the results were double checked, no obvious reason could be found to explain why the difference in pressure drops between a fuel bed with a height of 50 cm and a fuel bed with a height of 75 cm was considerably smaller than the difference between a 25 cm and 50 cm fuel bed. The following was concluded:

i) The pressure drop (cm H_2O/m) is about 5.75 times higher in a rice hull char bed than in an unburned rice hull bed. This ratio is independent of the height of the fuel bed.

ii) Pressure drops are automatically kept low because superficial velocities above 10 cm s^{-1} will disturb the fuel bed and interfere with the gasification process.

iii) The maximum pressure drop will be about 12 cm H_2O/m.

Theoretical Treatment of the Pressure Drop in a Rice Hull Fuel Bed

The pressure drop through a fixed bed can be predicted accurately if the input parameters are known with a certain precision. In general we may assume Δp (ε_β, ρ, v_s, L, γ, D_p). Ergun (1954) has proposed an equation to predict this pressure drop based on Δp (ε_β, ρ, v_s, L, D_p). He did not include the surface roughness.

Ergun equation: $$\frac{\lambda \varepsilon_\beta^{\ 3}}{1 - \varepsilon_\beta} = \frac{150\ (1 - \varepsilon_\beta)}{Re} + 1.75 \qquad (6\text{-}20)$$

where $\lambda = \dfrac{D_p}{\rho\ V_s^2}\ \dfrac{\Delta p}{L}$

$D_p = \dfrac{6\ V_p}{A_p}$

$Re = \dfrac{D_p\ \rho\ V_s}{\mu}$

V_p = volume of particle

A_p = surface area of particle

A review by McDonald et al (1978) of correlations for pressure drops through a fixed bed confirms the accuracy of the Ergun equation. The authors suggested that the additional parameter of surface roughness of the particles be incorporated into the equation. Their conclusions are given as follows:

$$\frac{\lambda \varepsilon_\beta^3}{1 - \varepsilon_\beta} = \frac{180\ (1 - \varepsilon_\beta)}{Re} + 1.8 \qquad \text{(smooth particles)} \qquad (6\text{-}21)$$

$$\frac{\lambda \varepsilon_\beta^3}{1 - \varepsilon_\beta} = \frac{180\ (1 - \varepsilon_\beta)}{Re} + 4.0 \qquad \text{(roughest particles)} \qquad (6\text{-}22)$$

Solving for the parameter Δp yields:

$$\Delta p = \left[\frac{150\ (1 - \varepsilon_\beta)}{Re} + 1.75\right] (1 - \varepsilon_\beta) \frac{\rho V_s^2 L}{D_p \varepsilon_\beta^3} \qquad \text{(Ergun)}$$

$$\Delta p = \left[\frac{180\ (1 - \varepsilon_\beta)}{Re} + 1.8\right] (1 - \varepsilon_\beta) \frac{\rho V_s^2 L}{D_p \varepsilon_\beta^3} \qquad \text{(smooth particle)}$$

$$\Delta p = \left[\frac{180\ (1 - \varepsilon_\beta)}{Re} + 4.0\right] (1 - \varepsilon_\beta) \frac{\rho V_s^2 L}{D_p \varepsilon_\beta^3} \qquad \text{(rough particle)}$$

Because it is difficult to draw a line between smooth and rough particles, the Ergun equation will be applied to the experimental data for a rice hull bed of 1 m height.

The Ergun equation indicates that Δp will be very sensitive to changes in D_p and ε_β, the physical parameters that can be measured with the least accuracy. We assume

$$D_p = \frac{(6)(4.75\ mm^3)}{(115.66\ mm^2)} = 0.246\ mm$$

where A_p and V_p were computed from the data given in Sections 6.2 and 6.3. The other physical constants are given as:

$$\varepsilon_\beta = 0.85, \; D_p = 2.46 \times 10^{-4} \text{ m}$$

$$\rho = 1.204 \text{ kg m}^{-3}, \; \mu = 181 \times 10^{-7} \text{ kg m}^{-1} \text{ s}^{-1} \text{ (at 15°C, 1 atm).}$$

The best fit for the experimental data was obtained by choosing D_p = 2.28×10^{-4} instead of $D_p = 2.46 \times 10^{-4}$ with a fixed $\varepsilon_\beta = 0.85$. The result is given in Figure 6-20.

It is noted that the Ergun equation overestimates the pressure drop by at most 15%. However this equation can not be used to predict pressure drops at superficial velocities higher than 21 cm s^{-1} because of disturbances in the fuel bed.

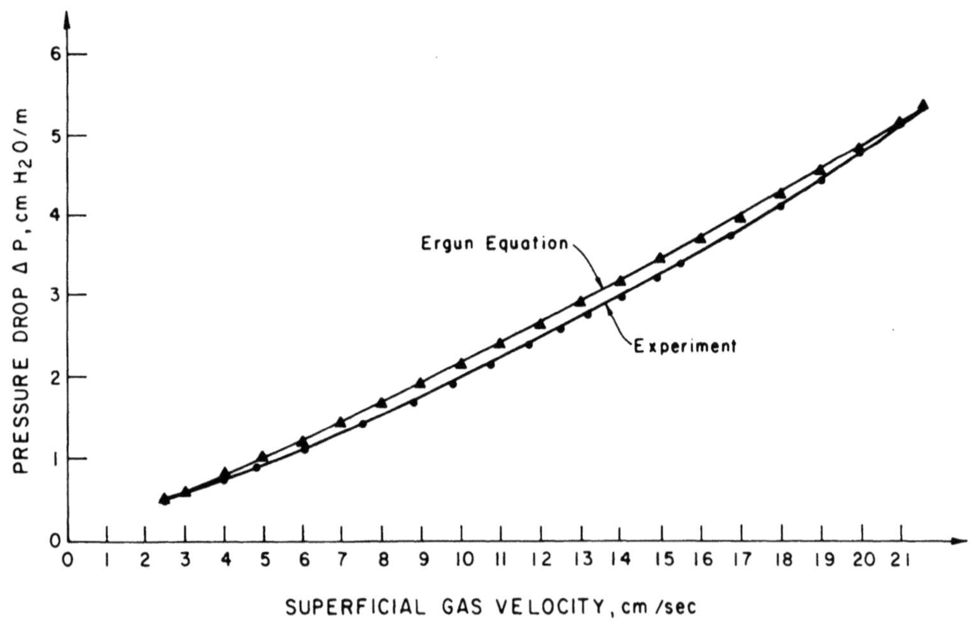

Figure 6-20. Pressure drop across a rice hull fuel bed, experimental and theoretical results.

Size Distribution of Rice Hulls and Rice Hull Char

Introduction

The high void space fraction in a rice hull bed, the large pore volume generated during gasification, and the minimal size reduction of rice hulls undergoing gasification make the residue of gasification (the char) a potentially efficient filter material for the raw gas. On the other hand, the friable consistency of rice hull char makes it difficult to handle without pulverizing a sizeable fraction. Pulverization may even occur in the lower layers of the gas producer, caused by interaction with the grate and by higher superficial velocities through the grate and gas exit in downdraft gas producers. The carryover of rice hull char has been a problem in the units tested. It is therefore of interest to establish a size distribution curve and to compute the extent to which a disengagement zone would separate fine solids (< 1 mm) from the gas stream. While these theoretical considerations will be discussed in Chapter 10, the size distribution of rice hulls and rice hull char is given here.

Size Distribution

The size distribution was determined using a set of Tyler wire mesh screens with square openings ranging from 0.05 mm to 2.5 mm. The screens were mounted in a standard screen shaker for the size distribution tests. It is noted that a particle falling through the wire mesh screen of size X during the shaking process was considered to be smaller than X. However, the particle is not necessarily smaller than X in all dimensions. Any rice hull remaining on the screen was considered to be larger than X. Thus it is possible that a particular particle could have fallen through given a different orientation of particle with the screen.

The results for rice hulls and char (obtained from a downdraft gasifier) are shown in Figures 6-21 and 6-22. Two types of char were tested. In one case the char was tested within hours after gasification while the second batch was tested after smoldering had ceased, 48 hours later. It is noted that when char from a rice hull gas producer is stored in a steel drum, char will smolder for extensive periods (days). Smoldering of the gasifier residue completes burnout (carbon removal). The silica skeleton formed is increasingly friable as the carbon content is reduced and any test performed will show considerable size reduction. With respect to the experiments, it is noted that:

a) Uncontrolled smoldering significantly increases the fine particle content of the char. For instance, the fraction of par-

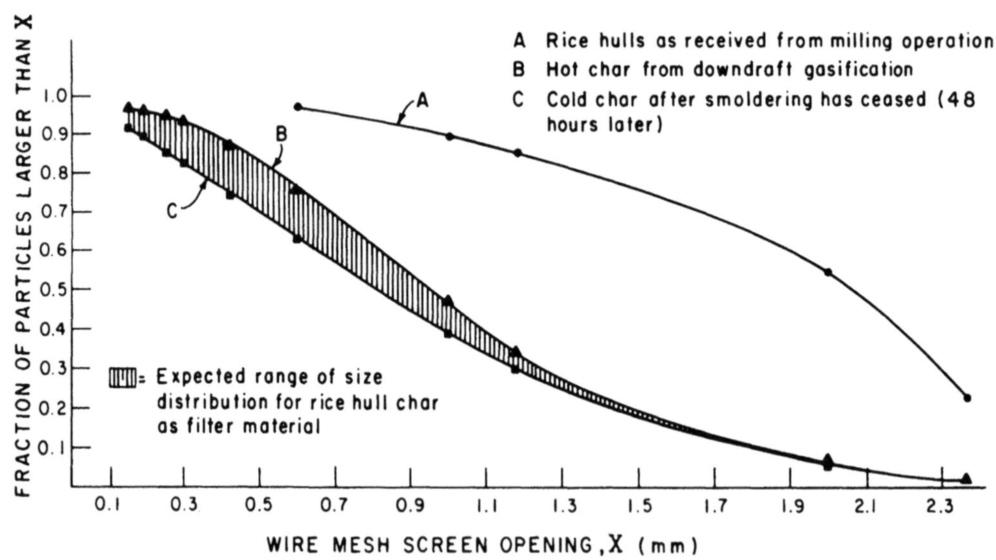

Figure 6-21. Size distribution of rice hulls and char removed from downdraft gasification.

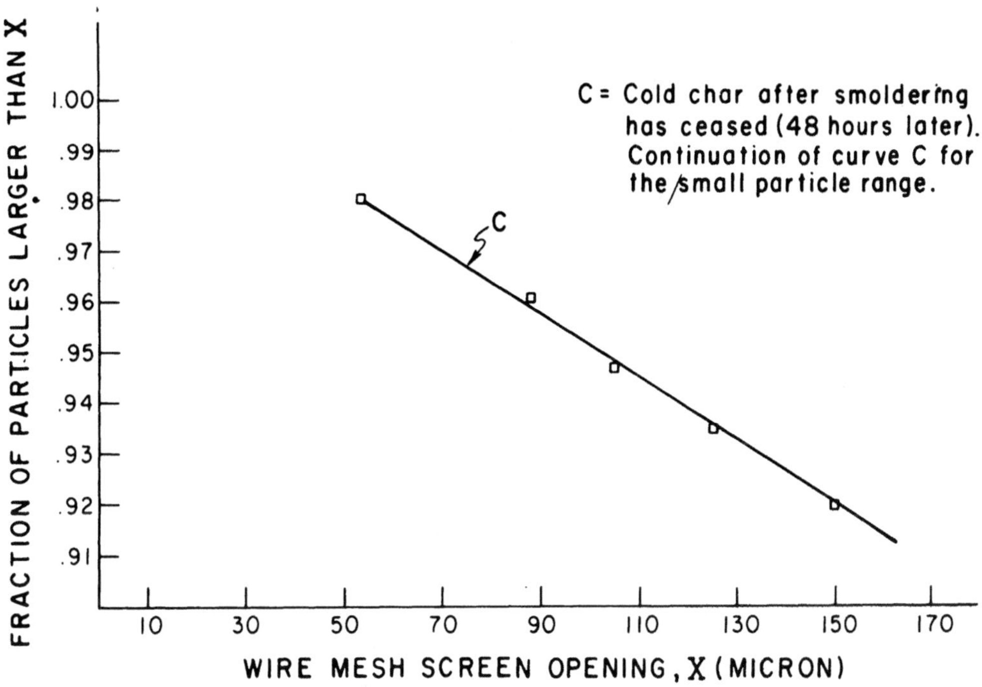

Figure 6-22. Fine particle content of rice hull char from downdraft gasification.

ticles smaller than 500 μm increases from 15% to 30%.

b) Particles smaller than 600 μm are mostly generated during the gasification process since only 2% were present before gasification as compared to 23%-36% afterwards.

c) The largest fractional increase is observed for the category of particles smaller than 1.5 mm. Initially, 25% of the particles fell into this category. This fraction increased to 82% after gasification.

d) The fraction of particles smaller than 50 μm is around 2% (particles smaller than 50 μm will be more difficult to remove from the gas stream in conventional cyclones and wet scrubbers).

6. LIST OF SYMBOLS

$A_{\beta\sigma}$	specific surface area of rice hulls (m^2/m^3)
A_p	surface area of single rice hull (m^2)
D_p	rice hull diameter (m)
L	length of rice hull column (m)
m	mass of rice hulls (kg)
m_s	mass of single rice hull (mg)
V	control volume (cm^3)
V_α	apparent volume of rice hulls (cm^3)
V_β	void space within the fuel bed (cm^3)
V_γ	pore volume of rice hulls (cm^3)
V_σ	volume of the solid matter of rice hulls (cm^3)
V_m	measured volume (cm^3)
V_p	volume of single rice hull (cm^3)
Re	Reynolds number
v_s	fluidizing velocity ($m\ s^{-1}$)
X	wire mesh screen opening (mm)
x_i	weight fractions of elements
γ	surface roughness
ρ	gas density ($kg\ m^{-3}$)
ρ_a	apparent density of rice hulls ($g\ cm^{-3}$)
ρ_b	bulk density of rice hulls ($g\ cm^{-3}$)
ρ_p	density of paraffin ($g\ cm^{-3}$)
ρ_t	true density of rice hulls ($g\ cm^{-3}$)
ρ_w	density of water ($g\ cm^{-3}$)
ε_i	phase fractions, where $i = \alpha, \beta, \gamma, \sigma$

ε_f porosity of rice hulls

ε_t total void space fraction of a rice hull fuel bed

ϕ equivalence ratio

Δp pressure drop across a rice hull bed (cm H_2O/m)

μ gas viscosity (kg m^{-1} s^{-1})

6. REFERENCES

1. Bartha, P. and A. Huppertz. 1974. The structure of silica in rice husks and their crystallization. Proceedings, Rice By-Product Utilization International Conference, Valencia, Spain.

2. Beagle, E. 1978. Rice husk conversion to energy. FAO Agricultural Services Bulletin #31.

3. Beagle, E. 1981. Personal Communication, Davis, California, USA.

4. Ergun, S. 1952. Fluid flow through packed columns. Chem. Eng. Prog. 48(89).

5. Goss, J.R. 1980. Personal Communication, Davis, California, USA.

6. Groeneveld, M. 1982. Personal Communication, Berlin, West Germany.

7. Houston, D.F. 1972. Rice chemistry and technology. Monograph Series IV, Chapter 12. American Cereal Chemist Association, St. Paul, MN.

8. Jenkins, B.M. 1980. Downdraft gasification characteristics of major California residue-derived fuels. PhD Thesis in Engineering, University of California, Davis.

9. Jones, J.D. 1953. New refractory from vegetable source. Canadian Metals. January.

10. Kaupp, A. and J.R. Goss. 1982. State-of-the-Art for small scale (to 50 kW) gas producer engine systems. U.S. Agency for International Development, Washington, D.C. 20523.

11. Levin, Ernest M. and Carl R. Robbins. 1964. Phase diagrams for ceramists. The American Ceramist Society, Columbus, Ohio.

12. Mantell, C.L. 1968. Carbon and Graphite Handbook. Inter-Science Publishers, New York.

13. McDonald, I.F. 1979. Flow through porous media - the Ergun equation revisited. Ind. Eng. Chem. Fundam. 8(3).

14. Nordstroem, O. 1964. Redogerelse for riksnamndens for ekonomisk forsvarsberedskap forsknings-och forsoksverksamhet pa gengasomradet vid statens maskinprovningar 1957-1963. Overstyrelsen for Ekonomiskt Forsvar, Box 7830, S-10398, Stockholm, Sweden.

15. Osman, E.A. 1982. A study of the effects of ash chemical composition and additives on fusion temperature in relation to slag

formation during gasification of biomass. PhD Thesis in Engineering, University of California, Davis.

16. Toropov, N.A. et al. 1972. Handbook of phase diagrams of silicate systems. Translated from Russian, Israel Program for Scientific Translations, Jerusalem.

17. Tucker, P.W. 1944. A study of the physical and chemical properties of rice hulls. M.S. Thesis, Louisiana State University, Department of Chemistry.

7. PHYSICAL APPEARANCE OF RICE HULLS UNDER THERMAL DECOMPOSITION

Introduction

During my first trial runs using rice hulls in standard wood gasifiers of the updraft and downdraft type, it became apparent that the quickly changing flow properties and the unique thermal decomposition characteristics of rice hulls cause severe operational problems in small gasifiers of 15 cm to 30 cm hearth diameter. For instance, common slot grates used in coal and wood gasification can neither remove the rice hull ash continuously nor prevent the rice hulls from falling through. Another extreme example of the variable nature of rice hull flow properties is shown in Figure 7-1. An experimental hopper with 70° and 45° walls was used to feed the rice hulls into a downdraft gas producer. The rice hulls were loosely poured into the hopper, but because of gravity feeding the rice hulls became compacted and formed a vertical wall during the trial run. Additional difficulties were encountered in updraft gasification. Tar vapors condensed in the cooler upper layers and formed a sticky coating of charred rice hulls and tarry liquids, as shown in Figure 7-2. The frequent flow distortion and channel generation within the fuel bed could not be attributed to one particular cause. The next step was to examine closely the physical appearance of rice hulls and rice hull pellets before and after gasification to see if any insights into fuel bed disturbance could be found.

Micrographs of Rice Hulls Before Thermal Decomposition

A schematic diagram of a single rice hull has already been shown in Figure 6-6. A micrograph picture is shown in Figure 7-3. The geometrically unique outer surface of rice hulls with its thousands (about

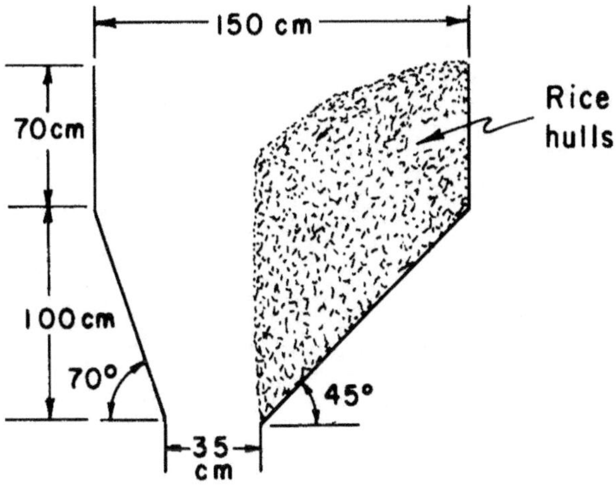

Figure 7-1. Vertical wall of rice hulls after one hour of operation.

Figure 7-2. Rice hull-tar conglomerate from updraft gasification.

Figure 7-3. Single rice hull before gasification (X 20).

Figure 7-4. Outer surface before gasification (X 180).

4000-5000) of bumps is shown in Figures 7-3 through 7-5. The inner surface of a single rice hull, on the other hand, is almost completely flat as shown in Figures 7-6 and 7-7. It is noted that even extensive scanning at a magnification of 6000X did not reveal any visible pores on the inner surface. Therefore pores on this surface must be smaller than 0.17 μm given that the human eye can detect 1 mm objects on the surface of the rice hull. One of the cross-sections used to determine the surface area of rice hulls (see Figure 6-7) is shown in Figure 7-8. A single bump on the outer surface is shown in Figure 7-9. It is again noted that rice hulls contain 15%-22% ash, most of which is SiO_2, as outlined in Chapter 6. Silicate systems have a marked effect on the stability of a rice hull undergoing thermal decomposition, as shown in the following micrographs.

Micrographs of Rice Hulls After Thermal Decomposition

The remarkable stability of a rice hull that has undergone thermal decomposition can be seen in Figure 7-10, which shows a single rice hull after complete combustion at 1200°C. All volatiles and carbon have been released leaving a snow-white silica skeleton that has the same shape and surface structure as the original rice hull. The slim appearance is due to the intense heat, which tends to curl the silica skeleton. The high temperatures have virtually no effect on the bumpy outer surface, as can be seen in Figure 7-11. Although temperatures in a gas producer are in the 900-1200°C range, snow-white residues shown in the figure are not found because the process operates at $\phi < 0.5$.

A typical outer surface of a gasified rice hull is shown in Figure 7-12. The appearance of a single bump after gasification is shown in

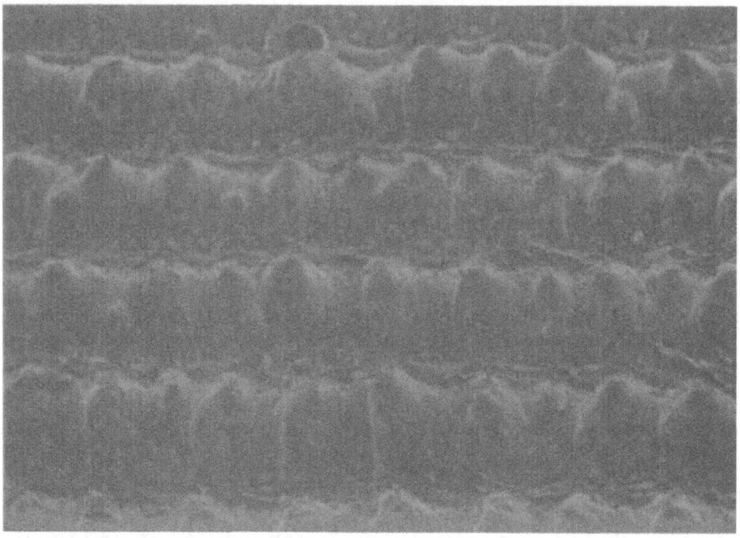

Figure 7-5. Outer surface before gasification (X 550).

Figure 7-6. Inner surface before gasification (X 500).

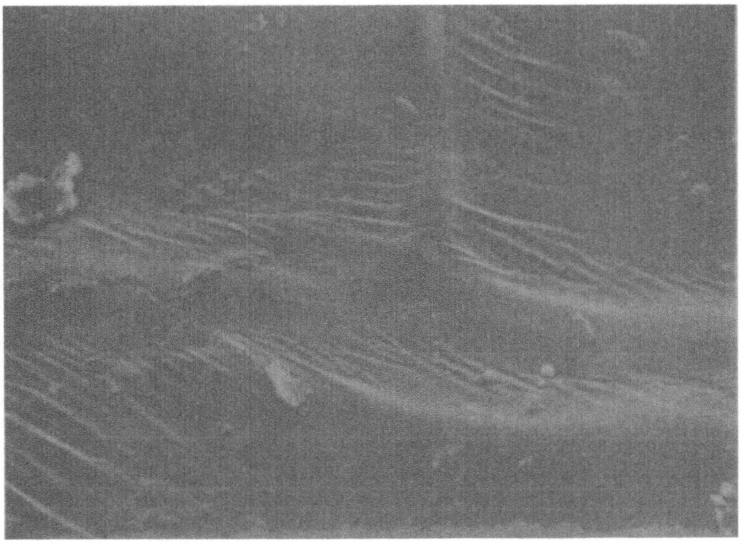

Figure 7-7. Inner surface before gasification (X 2000).

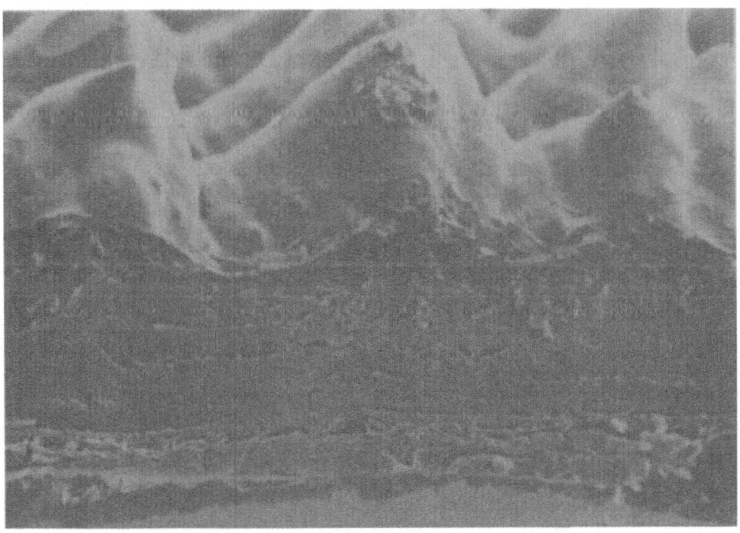

Figure 7-8. Cross-section of single rice hull before gasification (X 550).

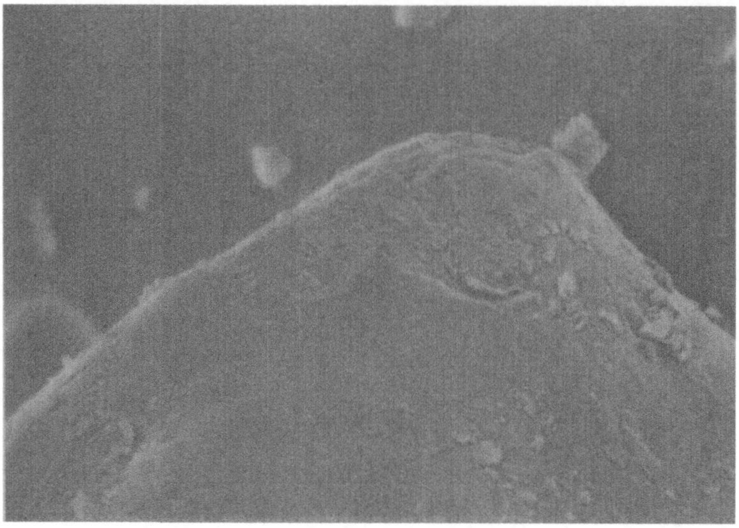

Figure 7-9. Bump on outer surface before gasification (X 2000).

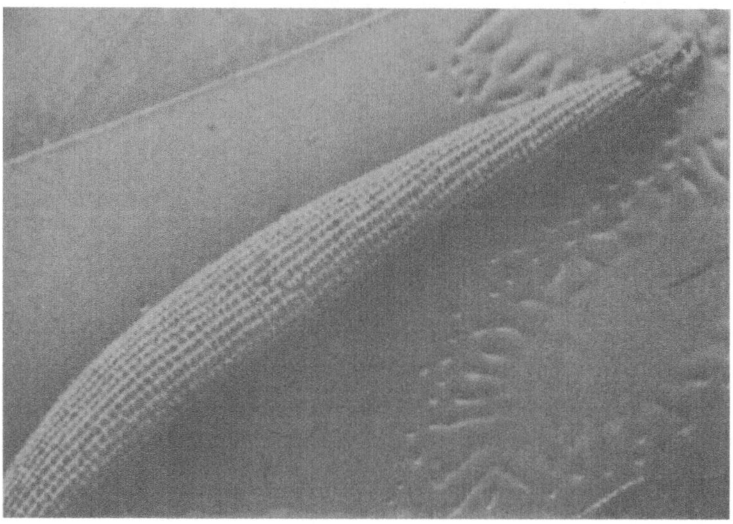

Figure 7-10. Rice hull after complete combustion at 1200°C (X 20).

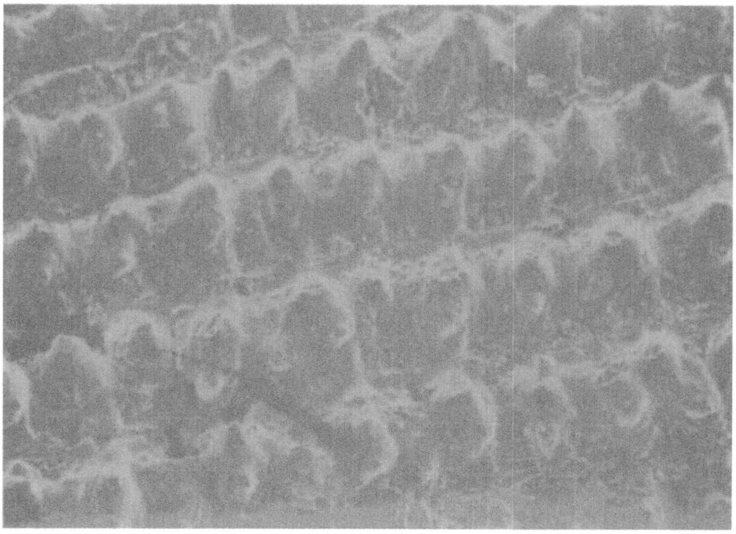

Figure 7-11. Rice hull after complete combustion at 1200°C (X 550).

Figure 7-12. Outer surface after gasification (X 550).

Figure 7-13. It is noted that the initially rounded bumps on the outer surface became sharp peaks after gasification. No pore formation was observed at this magnification. A completely different situation prevails along the inner surface of a rice hull, which is shown in Figure 7-14. From Figure 7-14 it can be concluded that the volatiles and carbon must be tightly woven into the silica skeleton of a rice hull. An indication as to the density of this network of pores can be seen from Figure 7-15.

Size Reduction of Pelletized Rice Hulls

Because of their stable silica skeleton, the size reduction of rice hulls under thermal decomposition is minimal. This is shown for pelleted rice hulls in Figure 7-16. In this case the remaining carbon in the pellets was completely burned off by long-term exposure (6 hours) to medium temperatures (900°C) in an atmosphere of air. The average measured spatial reductions are given in Table 7-1.

Table 7-1. Average Size Reduction of Rice Hull Pellets

	Before Gasification	After Combustion
Length (cm)	1.85	1.54
Diameter (cm)	0.95	0.77
Computed Volume (cm^3)	1.31	0.71

From Table 7-1 one may conclude that a 45% reduction in volume will occur for a single dimension reduction of 17%-19%. It is noted that depending on the gas producer and the ash removal design the char residue is brittle and disintegrates. For instance, in a downdraft or updraft gas producer the rice hull char will remain intact in most

Figure 7-13. Single bump after gasification (X 2000).

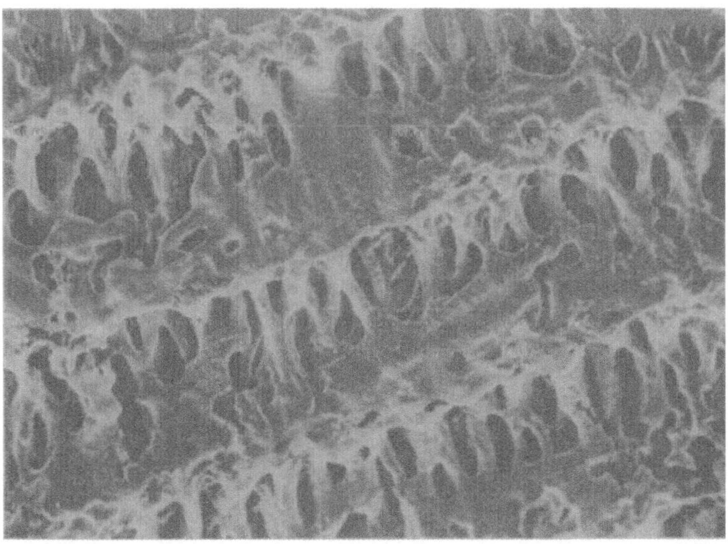

Figure 7-14. Inner surface of rice hull after gasification (X 550).

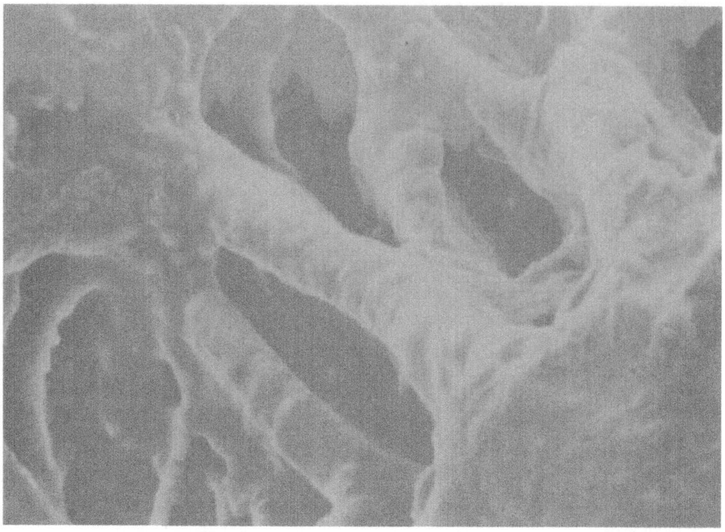

Figure 7-15. Inner surface silica skeleton of rice hull after gasification (X 2000).

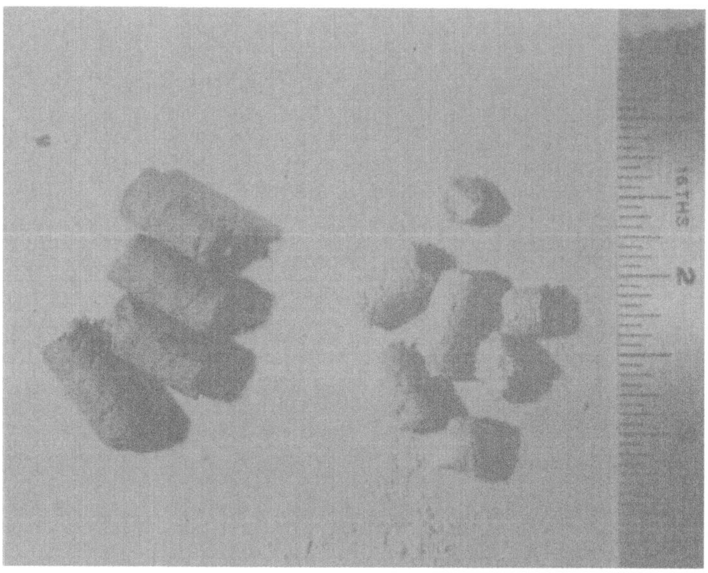

Figure 7-16. Size reduction of pelleted rice hulls under thermal decomposition. (Left: before, Right: after gasification).

cases. On the other hand, the use of sand as a heat transfer media (as in a fluidized bed gas producer) results in a fine dust-like residue.

Summary

It has been shown that rice hulls, because of their high ash content (in the form of a stable silica skeleton), exhibit unusual behavior during thermal decomposition. This behavior is not found in most other biomass fuels and is characterized below:

a) Minimal size reduction.

b) Stable silica skeleton is friable.

c) Large differences in the number of pores that open up along the inner and outer surfaces.

d) Highly porous char residue (see Figures 7-14 and 7-15).

8. LOW TEMPERATURE ENERGY CONVERSION OF RICE HULLS

Introduction

Combustion of a carbonaceous fuel includes a pyrolytic component. Pyrolysis takes place when the fuel is heated to a temperature between 250-500°C. At these temperatures, the carbonaceous fuel will begin releasing its so-called volatile components. The pyrolysis process can be characterized as the first stage of combustion. This stage takes place at relatively low temperatures of 250-500°C. Pyrolysis is not limited to solid fuels. A diffusion flame of producer gas from rice hull gasification and a schematic sketch of the flame is shown in Figures 8-1 and 8-2.

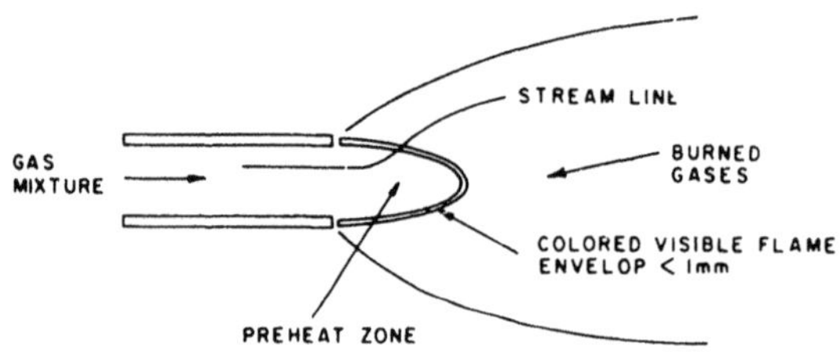

Figure 8-1. Schematic of producer gas diffusion flame.

In this case, oxygen is diffusing towards the very thin flame envelope from the outside. It will be consumed at the outer surface of the flame envelope. Consequently the inner preheat zone is the pyrolysis zone. In this zone low temperatures of 300-600°C are sufficient to induce the generation of pyrolysis products such as solid carbon (soot) which is in turn generated from the cracking of hydrocarbons in the gas. This process takes place within a fraction of a second in an oxygen-free atmosphere.

Figure 8-2. Orange diffusion flame from raw producer gas (solid particles, tar).

Rice hulls fed into a fluidized bed gas producer will be heated up within a fraction of a second to 500°C in an oxidizing atmosphere as shown in Section 8.3. Thus, although this process is completely different from that mentioned above, it also forms a similar range of products.

In updraft rice hull gas producers, the rice hulls move through a spatially fixed temperature gradient (see Figure 5-1). In both cases they will undergo a slow pyrolysis (on the order of minutes) in an oxygen-free atmosphere (< 2.5% oxygen). Therefore the following parameters characterize a pyrolysis process:

a) Temperature

b) Amount of time used to heat the reactants

c) Amount of oxygen present, which is used in a combustion mode to provide the heat necessary to sustain pyrolysis.

Products of Pyrolysis

The process can generate liquid, solid and gaseous products most of which are highly undesirable if the gas is used as an engine fuel. For instance, the separation of carbon and hydrogen in the pyrolysis zone of a diffusion flame results in solid carbon (soot). The presence of this solid carbon gives the flame a characteristic orange color.

The pyrolysis zone in a rice hull gas producer generates a considerable amount of liquid products. The mixture is called tar which is mainly higher hydrocarbons and water. A list of trace species that are also found in tar is given in Table 1-1. An important issue in the pyrolysis of coal or a biomass such as rice hulls is how to maximize the desired product(s). In general, converting a biomass solid fuel to a liquid or gaseous fuel is matter of adding hydrogen to the fuel. This can be seen from Table 8-1 which lists the C-H composition of some fuels.

Table 8-1. Fuels and their C-H Composition

Rice hulls	$CH_{0.7}$	(solid)
Coal	$CH_{0.8}$	(solid)
Water-free tar	$CH_{1.5}$	(liquid)
Light oil	$CH_{1.8}$	(liquid)
Gasoline	$CH_{2.25}$	(volatile liquid)
Natural gas	$CH_{3.9}$	(gas)

Two major pyrolysis processes have been classified according to their products. The following section discusses each of these often used processes:

a) Charcoal production

This ancient method was mainly used to generate a smokeless fuel with a high energy content. Thus the product of interest is the solid phase. Although charcoal production is mostly associated with wood, it can be done with most other biomass fuels as well, including rice hulls. Although large commercial charcoal production installations often use the gaseous phase to generate the necessary heat to devolatilize the biomass, it is difficult to find a use for the condensable phase. Heavy and light oils can only be marketed as fuels or engine oils if their viscosities and boiling point curves can be guaranteed within a narrow range. In addition a variety of corrosive acids are generated during pyrolysis and may harm burners or engines. Nevertheless the Italian rice hull gas producers operated prior to World War II generated a sizeable amount of heavy tar (asphalt) that was a very valuable byproduct at this time (see Section 10.2).

Almost all charcoal production equipment burns part of the fuel to generate the heat necessary for the pyrolysis process. As shown in Figure 5-4, the pyrolysis process is endothermic or at best thermally neutral. It is noted that optimization of the desired product, charred rice hulls, and minimization of the smokiness of the fuel cannot both be achieved. This problem becomes critical with a high ash fuel such as rice hulls. It will be shown later that the removal (using heat) of all nitrogen, hydrogen and oxygen in the fuel (the smoke generators) cannot be achieved without the removal of a considerable amount of carbon. A smokeless rice hull charcoal is therefore characterized by a high ash content. A large fraction of the initial energy stored in the fuel is lost to the gaseous and liquid phases, which are generally not utilized.

b) Tar production

There are a few situations in which the objective is to optimize tar production. In these cases the carbon ash residue (char) and the gases are only byproducts. Based on a very large amount of research done on the pyrolysis of coal, it is generally agreed that rapid heating increases the tar yield significantly and that at high heating rates and particles smaller than 50 μm, the particle size does not influence the tar yield (J.B. Howard 1981). Based on their results, the various rates of temperature increase during pyrolysis can be defined as follows:

Rapid (flash) pyrolysis	10^3 - 10^5°C s^{-1}
Intermediate pyrolysis	10 - 100°C s^{-1}
Slow pyrolysis	0.1 - 1°C s^{-1}

Rapid pyrolysis which optimizes the tar yield is therefore well achieved in a fluidized bed gas producer although tar production is certainly not the objective of such a unit (see Section 8.3). In order to market the liquid products, fractional distillation and removal of the acids in the tar is necessary. Thus utilization of the liquid phase is only feasible on a large scale.

In the gasification of rice hulls for engine fuel, the objective is to have as little tar in the gas as possible. However, for all types of gas producers one uses, tar will always be generated because pyrolysis will always be occurring. To reduce the amount of the tar in the raw gas three independent paths can be followed:

a) Very slow volatilization which may not decrease tar formation as much as desired.

b) Tar cracking within the gas producer (Chapter 9).

c) Pyrolysis of the rice hulls prior to their use as a gasifier fuel (Section 8.3).

Mechanism of Pyrolysis

The formation of volatile matter depends primarily on the hydrogen-to-carbon ratio of the fuel. Equations 5-10 and 5-11 are used to predict the anhydrous tar and water fractions of the pyrolytic products of coal for a 500°C assay. It is noted that the yields of the various products are very dependent on the type of laboratory assay used. The most commonly used tests for coal in the U.S. are the Fischer and Gray King assays. Both are carried out at up to 500-600°C or until the evolution of volatiles ceases. The devolatilization of coal has been described by Probstein (1982) as follows:

> "During active decompostion and temperatures around 500°C, it is the weak aliphatic links of the coal that are ruptured with subsequent release of the large molecular ring free radicals. The tar and light oils are formed from the stabilization of these free radical fragments by the hydrogen evolved".

This quote may serve as a reminder that the description of the pyrolysis process in terms of reversible kinetics or even in terms of the reactions described in Equations 5-1 through 5-9 has resulted in false conclusions concerning the overall gasification process.

No governing theory exists for the pryolysis process. In fact, the uncertainty about the process is so great that there is disagreement as to whether the product yields change with heating rate (Anthony, Howard 1976 and Howard 1981). The main obstacle to a unified theory is the difficulty in distinguishing between primary and secondary reactions during the process.

The primary reactions are those involving the breakdown of the hydrogen, oxygen, hydroxyl and nitrogen bonds in the fuel. Once this stage is reached, gaseous and liquid products are generated that will undergo homogeneous and heterogeneous reactions, which are secondary reactions. These secondary reactions give a false impression as to the importance of the highly irreversible primary reactions. It will be shown in Chapter 9 that high temperatures in the gasifier alone can catalyze the conversion of all tarry liquids into gases. Thus the primary pyrolysis products have therefore been converted into an equilibrium mixture of gases and liquids whose composition is dependent on temperature and pressure.

In order to accurately assess the gasification process, a clear distinction must be made between the products of pyrolytic reactions and the products of gasification reactions. To demonstrate this, the methane generation is used as an example. It has been shown in Chapter 5 that methane as well as the gaseous higher hydrocarbons (C_2H_m, C_3H_m and C_4H_m) are present only in small amounts (< 1%) in the equilibrium compositions. On the other hand, methane is found in large quantities (3% - 12%) in the reported dry gas compositions from downdraft, updraft and fluidized bed gas producers. The fact that these high methane contents are primarily caused in the pyrolysis zone and that the methane content remains relatively stable at high temperature (≈ 1000°C) is shown in Figure 8-3, reported by Goodman et al (1958). The interesting feature of this figure is that the methane content increases with increasing temperature up to 600°C and that it remains relatively stable up to a temperature of 1000°C. These findings do not contradict the well-known observation that the methane fraction generated through reversible reactions such as 5-5 and 5-8 will decrease with temperature and becomes negligible above 800°C (Chapter 5). Instead, Goodman's results indicate the difficulty in finding a model for the pyrolysis process.

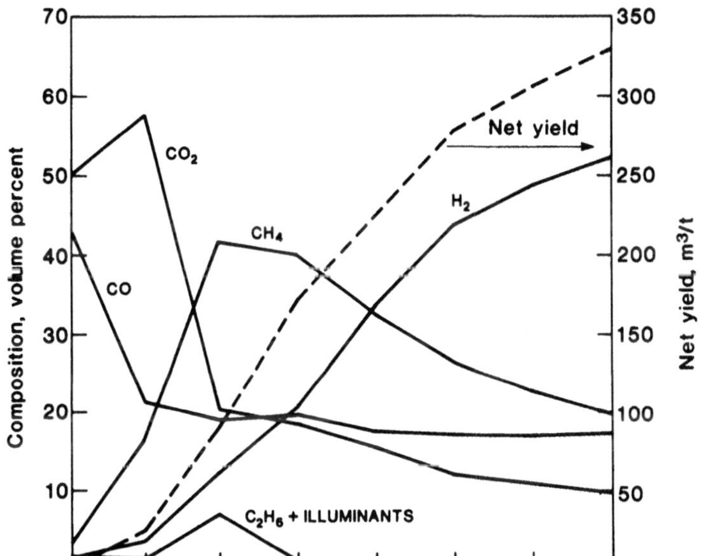

Figure 8-3. Composition of pyrolysis gas on a nitrogen and oxygen free basis from a sub-bituminous B coal (Probstein, 1982).

In particular, tar formation and tar yield in a gas producer must be explored further not just for operational reasons, but also because of the significant health problems associated with the compounds that make up tar. It is well known that polycylic hydrocarbons (PAH's) which are products of pyrolysis have an exceptionally tumorigenic potential (Harvey 1982). In particular, DMBA, shown in Figure 8-4 is one of the most potent carcinogens known. Harvey reports that the problem was identified as far back as 200 years ago by a London surgeon, Percival Pott, who related the soot in chimneys to the high incidence of epithelioma of the scrotum among chimney sweeps.

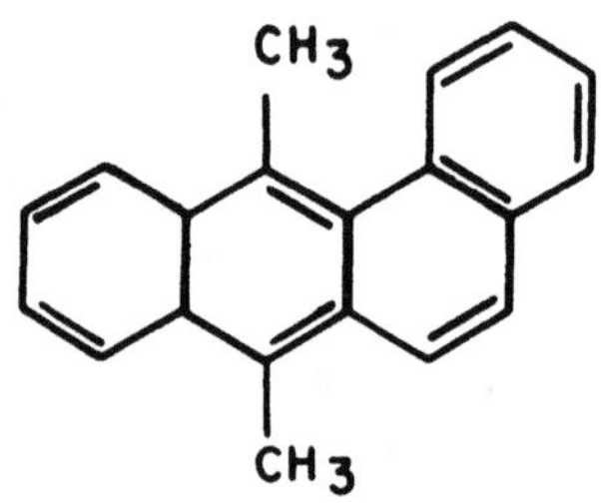

7, 12-DIMETHYLBENZ [a] ANTHRACENE
(DMBA)

Figure 8-4. DMBA.

In order to obtain some data on the pyrolysis of rice hulls and on the products of rice hull pyrolysis, laboratory experiments were performed in a nitrogen atmosphere. Charring of rice hulls was also performed using slight oxidation.

Pyrolysis Experiments in a Pure Nitrogen Atmosphere

The objective was to observe the formation of the solid, gaseous and liquid products as a function of temperature. Therefore a constant heat rate and an oxygen-free atmosphere were used for all tests. Because the literature is full of many definitions especially concerning distinctions between the phases, a set of definitions are presented:

Defintions

a) Volatiles: All gaseous and liquid products evolved up to temperature T.

b) Tar: The anhydrous fraction of the liquid products.

c) Tarry liquids: The entire liquid phase.

d) Fuel carbon: The amount of carbon that was initially present in the fuel as measured by the elemental chemical ultimate analysis (% weight, dry basis).

e) Char: The solid phase (residue) after pyrolysis.

f) Base carbon: The carbon left (fixed) in the char.

g) Volatile carbon: The carbon that has been evolved during pyrolysis.

h) Ash: The mineral fraction of the fuel or char.

i) Weight fraction consumed: The fraction of each of C, H, O, and N in the liquid and gas phases after pyrolysis.

The heat-up period for rice hulls is an important parameter with respect to tar yield and is computed in the following section.

Heat-up Period for a Single Rice Hull

The heat-up period for a single rice hull can be closely approximated by a nontransient solution of the energy equation. This assumption is justified if one can show that the resistance to heat transfer within the hull is negligible as compared to the resistance within the film. This means $N_{Bi} = \frac{hb}{2k} << 1$. Under these assumptions the dimensionless normed temperature function $\Theta(t) = \frac{\langle T \rangle - T_1}{T_o - T_1} = e^{-\frac{\langle h \rangle A}{\rho c_p V} t}$ describes the heat-up process.

$\langle T \rangle$ = average rice hull temperature

T_1 = temperature of the environment the rice hull is exposed to

T_o = initial temperature of the rice hull

$\langle h \rangle$ = surface averaged film heat transfer coefficient

ρV = weight of single rice hull

A = surface area of a rice hull

c_p = specific heat of a rice hull

b = thickness of rice hull

k = thermal conductivity.

One of the difficulties involved in this computation is in obtaining a close estimate of the film heat transfer coefficient, $\langle h \rangle$, which appears in the Biot number N_{Bi} and the temperature function. A review of experimental heat transfer data which has been compiled by Whitaker (1972) has been used to estimate $\langle h \rangle$. The particle size diameter is given by:

$$D_p = \frac{6V_p}{A_p} = \frac{(6)(4.75)\ mm^3}{(115.66)\ mm^2} = 0.246\ mm$$

where the volume and surface area have been measured as outlined in Chapter 6. The characteristic length L* is then given as

$$L^* = D_p\ (\frac{\varepsilon_\beta}{1-\varepsilon_\beta}) = (0.246)\ (\frac{0.85}{1-0.85}) = 1.394\ mm$$

Depending on the required gas quality a 15 cm diameter rice hull gas producer can generate between 2 and 10 Nm^3/h of producer gas. Therefore the real gas velocity u* is given by $u^* = \frac{Q}{\varepsilon_\beta A} = 0.185\ m\ s^{-1}$

assuming $Q = 10\ Nm^3/h$

$\varepsilon_\beta = 0.85$

$A = 0.01767\ m^2$

we have

$$N_{Re} = \frac{\rho u^* L^*}{\mu} = 2.13$$

assuming $\rho = 0.33\ kg\ m^{-3}$

$\mu = 400 \times 10^{-7}\ kg\ m^{-1}\ s^{-1}$

Correlations given by Whitaker (1972) do not cover this unusual combination of high void space fraction and low Reynolds number. One may conclude from Figures 6 and 7 of Reference 7 that $N_{Nu} < 2$ is justified, and therefore

$$h = \frac{2k}{L^*} = 249\ J\ m^{-1}\ s^{-1}\ K^{-1}$$

$$k = 0.1737\ J\ s^{-1}\ m^{-1}\ K^{-1}$$

$$\Theta(t) = e^{-\frac{\langle h \rangle A}{\rho c_p V} t} = e^{-4.45t}$$

$c_p = 2094\ J\ kg^{-1}\ {}^{\circ}C^{-1}$ (average for wood, not available for rice hulls).

$\rho V = 3.087 \times 10^{-6}$ kg.

The Biot Number is then given by $N_{Bi} = \frac{(249)(85 \times 10^{-6})}{(2)(0.1737)} = 0.061.$

This means the assumption of a nontransient solution of the energy equation is justified.

Example: A single rice hull exposed to a hot glowing carbon bed at 1000°C will heat up to 990° from an initial temperature of 20°C within 0.5 seconds. It is noted that the uncertainty lies with <h>. Varying the Nusselt number from 1 - 10 changes the calculated heat-up period from 1 to 0.1 seconds.

It is therefore concluded that fluidized bed gasification induces a rapid pyrolysis process while the following experiments simulate a slow pyrolysis process at heating rates of 0.3 - 0.4°C s^{-1}. The later heat rates are common in large-scale updraft and downdraft gas producers.

Experimental Set Up and Procedures

The experiments were conducted in a tube furnace connected to a tar and water sampling train as shown in Figure 8-5. Tar and water were

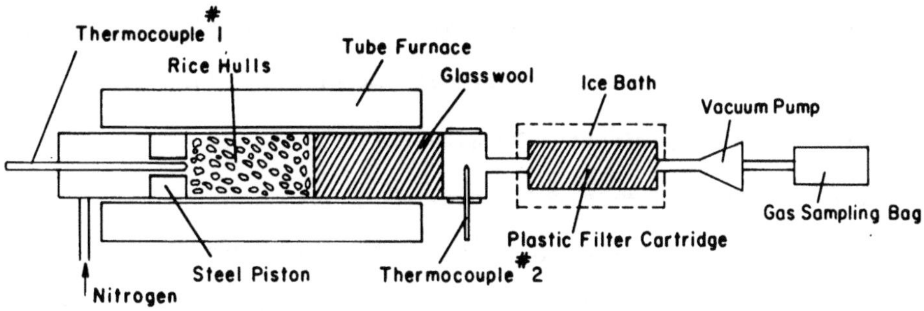

Figure 8-5. Rice hull pyrolysis apparatus.

filtered and separated by using a temperature resistant glass wool arrangement. Condensation and separation of tarry liquids usually results in inacurracies because separation is not complete. Thermocouple #1 indicated the final temperature of the experiment while thermocouple #2 (in conjunction with a heating collar) was used to keep the cooler end of the tar trap at 105°C to avoid condensation of water before it reached the ice bath. The pyrolysis experimental procedure was:

a) Loading of the tube with 4-5 gram of oven-dry rice hulls.

b) Purging of the entire system with nitrogen.

c) Generation of a vacuum (25 cm H_2O) to draw vapors and gases from the sample into the filter train and gas bags.

d) Heat-up rate of 0.35°C s^{-1} until the sample reached the desired temperature.

e) Rapid cooling of the tube with water after the desired temperature was reached.

f) Immediate sealing of all liquid and solid products in zip bags to avoid evaporation.

In some of the low temperature experiments (250-300°C) small amounts of light oils were carried over into the water collection system.

Discussion of the Experimental Results

The following data were collected:

a) Composition of the gas phase at 250, 300, 350, 400 and 450°C.

b) Ultimate elemental analysis of the rice hulls and of the solid residue of pyrolysis at 250, 300, 350, 400 and 450°C.

c) Weight fractions of the products of pyrolysis (char, gas, tar and water).

d) Weight fractions consumed of the elements C, H, N, O during pyrolysis.

e) Higher heating value of rice hulls and the char at 250, 350, 400 and 450°C.

f) Cumulative mass loss during pyrolysis.

Composition of the Gas Phase

The composition of the gas phase as a function of temperature is shown in Figure 8-6. Gas samples were tested for O_2, CO, CO_2, CH_4, C_2H_4, C_2H_6 and H_2. Hydrogen and C_2H_m were only found in very small amounts (< 1%) in the gas. The low hydrogen fraction indicates that there is not enough hydrogen present during low temperature pyrolysis to stabilize the radicals and prevent caking of the rice hulls.

As expected, the methane content was between 1% and 4% and did not increase with decreasing temperature. The predominant gaseous products were CO_2 and CO. No significant change in the CO content could be detected while a slight decrease in CO_2 occurred with increasing temperature.

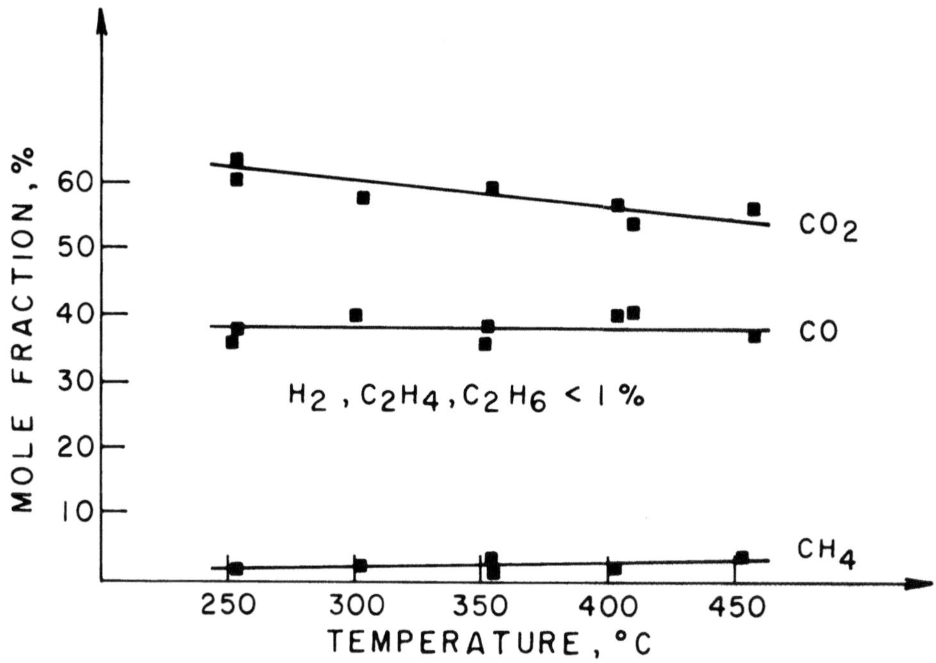

Figure 8-6. Dry gas composition as a function of temperature.

Ultimate Elemental Analysis of Rice Hulls and Char as a Function of Temperature

Charring of rice hulls is a wasteful process with regard to the fuel energy lost. Its only purpose is to generate a more reactive and a less smoky fuel. As outlined in Chapter 6, in the special case of the gasification of rice hulls in small gas producers, it may also prevent the caking of rice hulls during thermal decomposition at low temperatures. The object is therefore to remove as many of the tar forming elements (H, O, and N) as possible while preserving most of the carbon in the fuel. Release of volatiles and the retention of carbon cannot be simultaneously optimized. This is shown in Figure 8-7. Because the

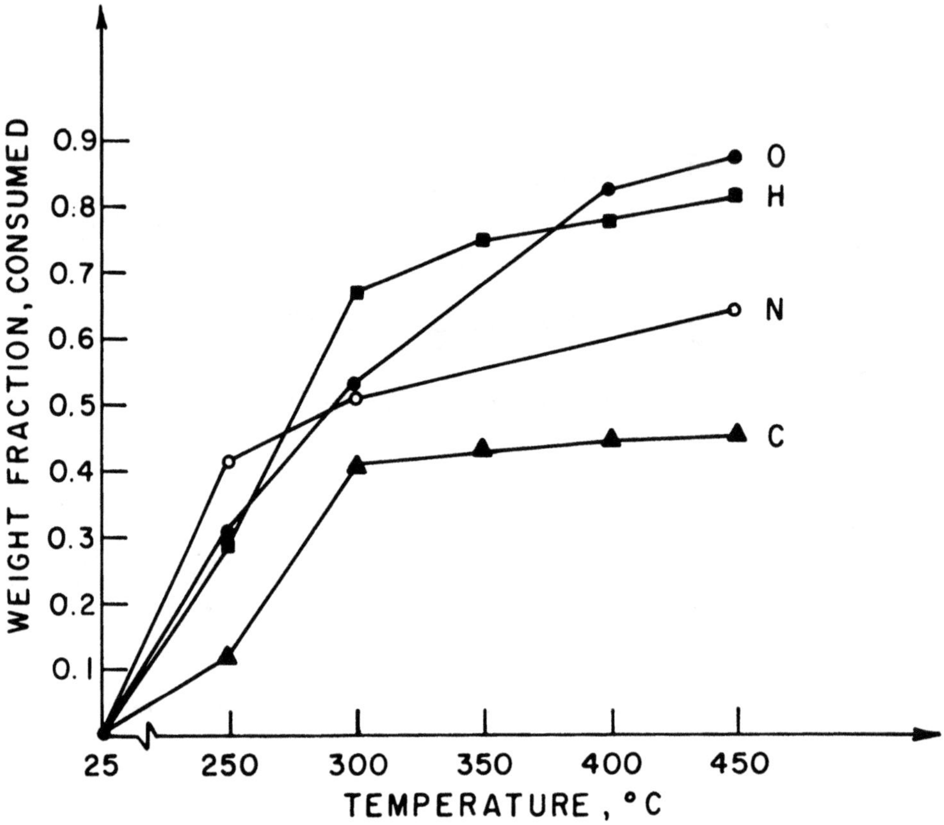

Figure 8-7. Weight fraction of C, H, N, O consumed as a function of temperature.

tests were conducted in a nitrogen atmosphere the release of volatiles becomes insignificant above 450°C. However, at this temperature only 87% of the oxygen, 82% of the hydrogen, and 64% of the nitrogen in the fuel were released along with 45% of the carbon. It is noted that the release of carbon between 300 and 450°C is insignificant, whereas large fractions of O, H and N are released in this temperature interval. If a smokeless fuel is desired it is advisable to char the rice hulls at a higher temperature, especially since the carbon loss at 450°C is not much greater than the carbon loss at 300°C. These results may not be used to predict the results of large-scale rice hull charring, because partial burning of a layer of rice hulls is allowed.

The outcome of any pyrolysis process in which some oxygen is present is heavily dependent on the length of time it takes to complete the process. An example is that of gasified black rice hull char at 200 - 300°C which, when exposed to ambient air, turns gray and completely changes its size distribution (Section 6.7) within 24 hours.

Weight Fractions of Char, Gas, Tar and Water

The results are shown in Figure 8-8. It is noted that no significant change is observed in the temperature interval of 350°C - 450°C, thus the bulk of the volatile matter is released below 350°C. Because of the heating rate of 0.3 - 0.4°C s^{-1}, the process is classified as slow pyrolysis which yielded 45% char and 37% tarry liquids which were separated into 10% tar and 27% water. The balance was gases mainly CO and CO_2. It is noted that in an ideal downdraft gas producer, all of the tar would be converted into gases while a large fraction of the water would not be decomposed, because the water-gas reaction is far from equilibrium.

In an updraft gas producer, which usually generates a gas with a 0% - 2% oxygen content (unreacted O_2 in the air), devolatilization takes place in an almost inert atmosphere. Figure 8-8 indicates why updraft gas producers generate excessive amounts of tarry liquids. These liquids necessitate extensive cleaning of the gas.

Model I introduced in Chapter 5 cannot be used to predict the yields of pyrolytic processes as outlined. The product distribution based on Model I is shown in Figure 8-9. It is noted the gas phase fraction is overestimated while the liquid phase fraction is underestimated. The model therefore fails to take into account the generation of tar which is not readily converted into gases.

Energy Balance

It is recalled that pyrolysis or charring of a biomass fuel has three main objectives:

a) Production of a less smoky, clean burning fuel that does not produce tar.

b) Production of a fuel with a higher heating value than the original feedstock it was made from.

c) Production of a more reactive fuel.

On the other hand the generation of charred biomass has the following drawbacks:

a) A considerable fraction of the energy in the initial feedstock is lost in the gas and liquid products.

b) The charred product is more delicate to handle and easily breaks up, such as soft wood charcoal or rice hull char.

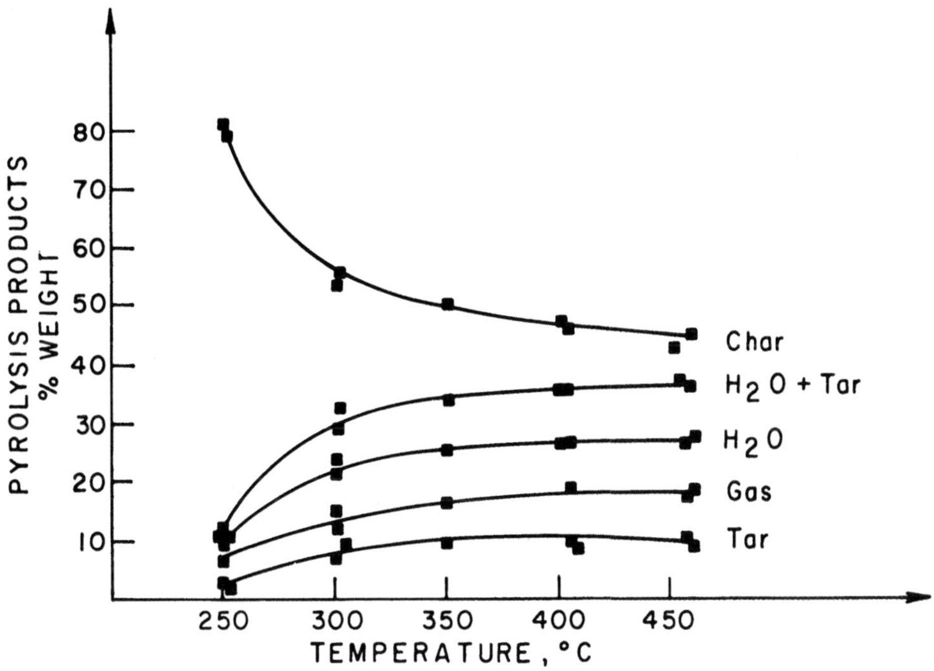

Figure 8-8. Weight fraction of the pyrolysis products as a function of temperature.

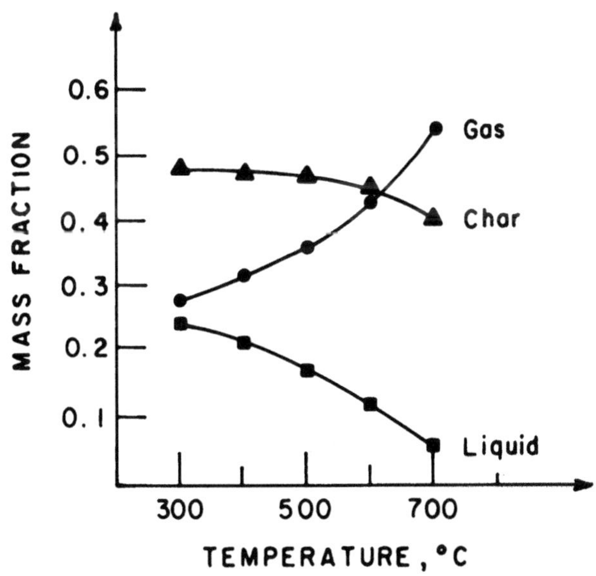

Figure 8-9. Weight fractions of the products from Model I at $\phi = 0$.

To show the striking difference between wood charcoal production and rice hull char production, the experimental results are compared to results from pyrolysis of pine sawdust and bark at 400°C in Table 8-2. From Table 8-2 it is obvious that rice hull pyrolysis does not generate the desired products. The charring process does not result in a product with a much higher energy content.

Table 8-2. Ultimate Analysis of Pine Sawdust + Bark, Charcoal and Rice Hulls and Char at 400°C

Specie	Wood*	Charcoal	Rice Hulls	Char
		% weight, dry basis		
Carbon	52.3	75.3	39.0	43
Hydrogen	5.8	3.8	4.9	2.2
Oxygen	38.8	15.2	33.9	11.2
Nitrogen	0.2	0.8	0.5	0.3
Sulfur	0.0	0.0	< 0.01	< 0.01
Minerals	1.4	3.4	21.6	43.3
HHV MJ kg^{-1}	19.7	27.1	15.3	16.3

*Data from Tillman (1978)

It is noted that the reason for the very unfavorable results is the initial high ash content of the feedstock. This yields a product of very little economical value. In this context it is of interest to look at the higher heating value of the charred rice hulls as a function of pyrolysis temperature as shown in Figure 8-10. From Figure 8-10 it can be seen that only charring at very low temperature of 250° will result in a reasonable energy loss of 15%. However, low temperature charring (< 250°C) does drive off an insignificant amount of volatiles and makes

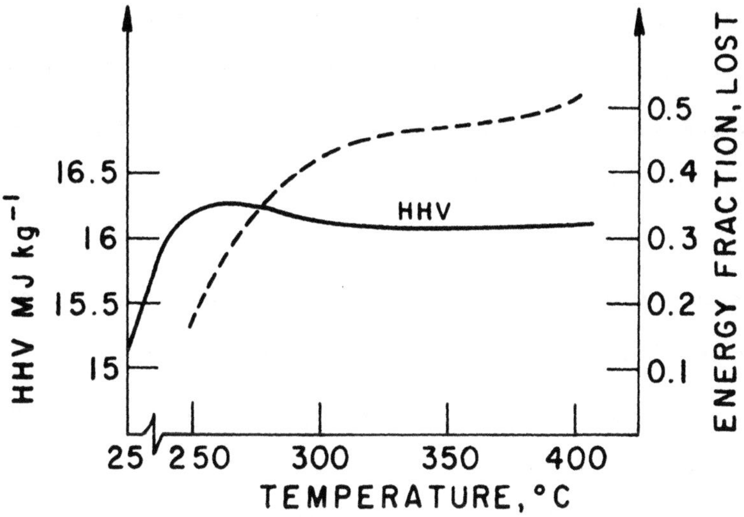

Figure 8-10. Higher heating value of rice hulls (A) and char and energy lost (B) during the pyrolysis process.

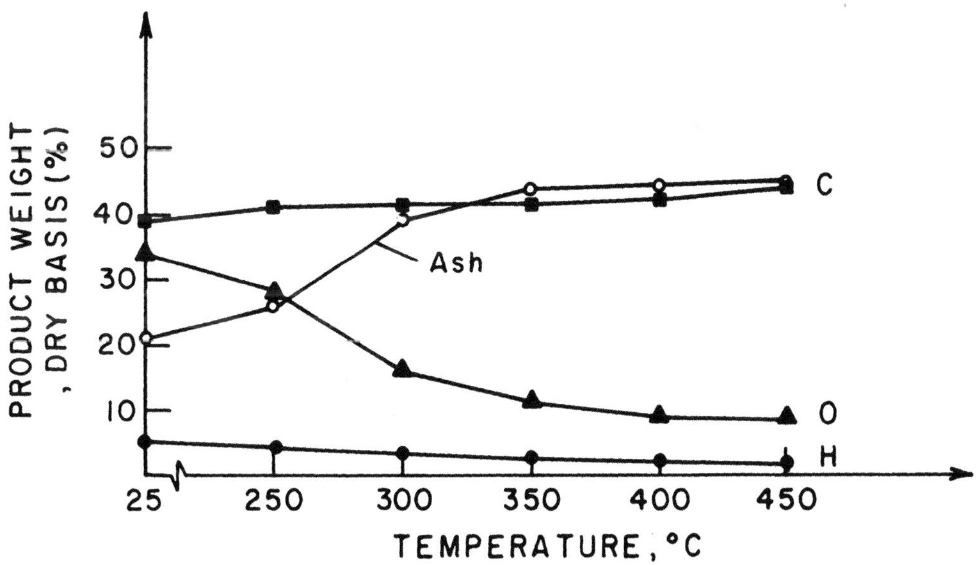

Figure 8-11. Weight fractions of ash, carbon and volatiles (H, O, N) in unburned rice hulls and rice hull char, as a function of temperature.

the process impractical. This is shown in Figure 8-11. For instance, slight charring at 250°C results in a product of 40% carbon and 60% volatiles and ash, while charring at 450°C will yield a product of 45% carbon, 45% ash and 10% volatiles.

Summary

The slow pyrolysis of rice hulls has been investigated at temperatures of 350°C to 450°C. The primary results are the following:

a) An average yield of 10% dry tar, 27% water, 18% gas and 45% char.

b) The charred rice hulls have a slightly higher heating value of 16 MJ kg^{-1} compared to 15.3 MJ kg^{-1} for rice hulls.

c) The energy fraction lost due to charring amounts to 45-55% on a dry basis.

d) The rice hull char consists of approximately 45% ash, 45% carbon and 10% remaining volatiles.

8. REFERENCES

1. Anthony, D.B. and J.B. Howard. 1976. Coal devolatilization and hydrogasification. American Institute of Chemical Engineers Journal 22.

2. Goodman, J.B. et al. 1953. Low temperature carbonization assay of coal in a precision laboratory apparatus. U.S. Bureau of Mines Bulletin 530, Government Printing Office, Washington, D.C.

3. Harvey, R.G. 1982. Polycyclic hydrocarbons and cancer. American Scientist. July-August.

4. Howard, J.B. 1981. Fundamentals of Coal Pyrolysis and Hydropyrolysis in Chemistry of Coal Utilization. Second Supplementary Volume. Wiley and Sons. New York.

5. Probstein, R.F. 1982. Synthetic Fuels. McGraw-Hill Book Company. New York.

6. Tillman, D.A. 1978. Wood as an Energy Resource. Academic Press. New York.

7. Whitaker, S. 1972. Forced convection heat transfer correlations for flow in pipes, past flat plates, single cylinders, single spheres, and for flow in packed beds and tube bundles. AICHE Journal 18(2).

9. TAR CRACKING IN A RICE HULL AND RICE HULL PELLET FUEL BED

Introduction

The renewed interest in small gas producer-engine systems (5-100 hp) has mostly involved downdraft (co-current) gas producers because of their ability to convert condensable pyrolytic products into noncondensable combustible gases. The evolution of gas producers for use with internal combustion engines has taken place as shown in Figures 9-1 through 9-3. The trend toward downdraft designs is due to the difficulties that occur in cleaning the highly contaminated (tar and water) gas generated in an updraft gas producer. However, the myth that downdraft gas producers generate a "tar-free gas" has never been proven and the fact is that all known operational downdraft and updraft gas producer-engine systems are equipped with extensive gas cleaning trains for tarry liquids and solid contaminants. The notion of a tar-free gas has been persistently associated with downdraft gas producers in scientific literature. It is therefore of interest to:

a) Analyse the reasons why downdraft gas producers could potentially generate a tar-free gas.

b) Determine why this process is difficult to control.

Tar Conversion in a Downdraft Gas Producer

Measurement Techniques

It was stated in Chapter 8 that the tarry vapors (which may be a mixture of higher hydrocarbons and water in mist form) have a wide range of boiling points, ranging from 100°C - 400°C. Large quantities of fine char particles in the exit stream (700 mg/Nm^3 - 6000 mg/Nm^3) act as

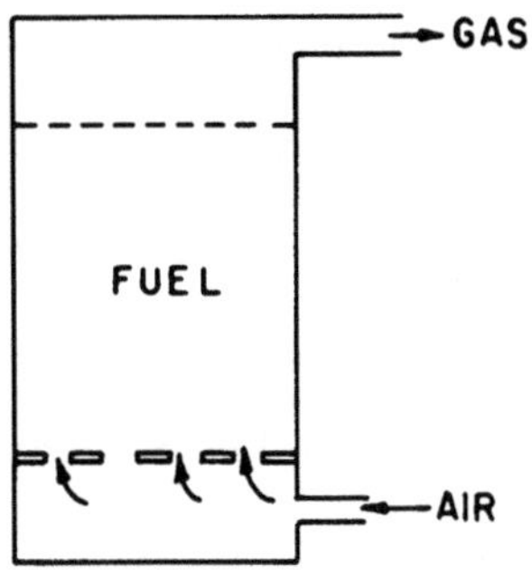

Figure 9-1

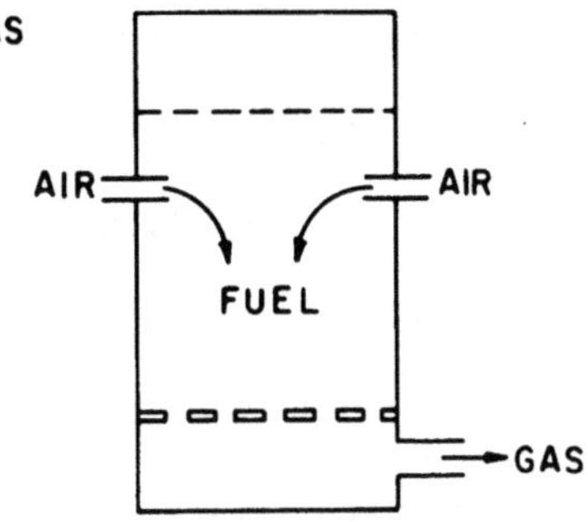

Figure 9-2

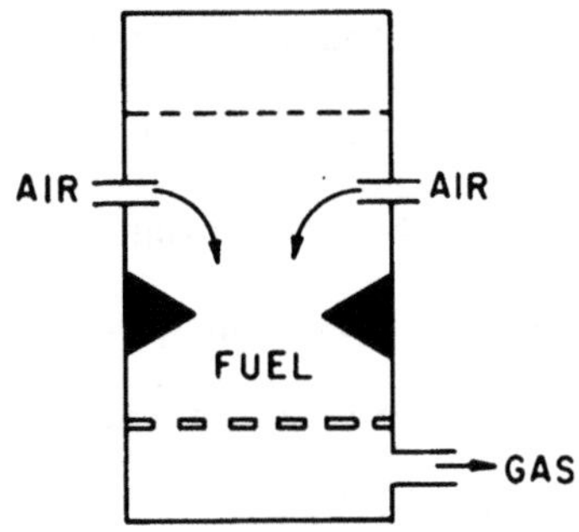

Figure 9-3

Evolution of gas producers for use with internal combustion engines.

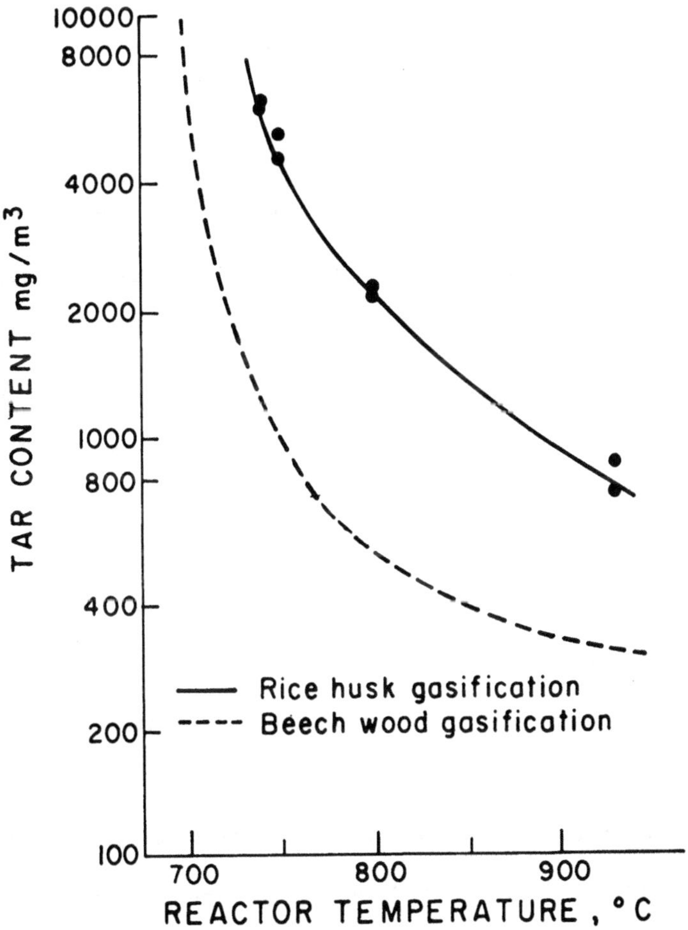

Figure 9-4. Tar content for beech wood and rice hull producer gas (Van den Aarsen, 1982).

nuclesation sites for tarry vapors. The first stage in processing the raw gas was usually to cool it at the gas producer exit in an ice bath. Thus, the resulting condensate is a tarry liquid that has already lost part of its high temperature fraction and contains solid carbon. This product is called the tar content of the raw gas and is usually measured in mg Nm^{-3} of dry gas. A typical result is given by Van den Aarsen (1982) in Figure 9-4 for beech wood and rice hulls. These experiments have been performed in a 30 cm diameter fluidized bed gas producer. The reported data indicate that the tar content of the raw gas is strongly dependent on the temperature.

Groeneveld (1980), in a study of a downdraft gas producer with a center tuyere, points out that "tar-free" means a tar content of less than 200 mg Nm^{-3} of dry gas. Based on the stringent requirements concerning maximum contamination in producer gas that is to be used in engines (less than 10 mg of total contaminants per Nm^3 of producer gas), it seems inappropriate to speak of "tar-free" gas in relation to a downdraft gas producer. However, use of the term "tar-free" is justified for producer gas obtained from charred fuels. In an ideally carbonized fuel, all the volatiles have been driven off. Only carbon and minerals remain as residue. As shown in Chapter 8 a highly carbonized fuel cannot be economically produced from rice hulls.

Mechanism of Tar Conversion

In general, there are three ways in which large hydrocarbon molecules can be broken into smaller ones in a downdraft gas producer.

a) Thermal decomposition, i.e. applying heat to the tar products in an oxygen-free atmosphere.

b) Thermochemical decomposition, i.e. applying heat to the tar in the presence of a catalytic surface. In practice, this involves utilization of the hot carbon-mineral surfaces available in a gas producer.

c) Combustion of tar with an oxygen carrier such as air.

All three mechanisms will contribute to tar conversion as follows:

a) It will be shown later that high temperatures provide enough energy to split higher hydrocarbons in the absence of any carbon-mineral surface.

b) In the presence of a carbon-mineral surface the tar cracking process consumes carbon. This reaction may be represented as:

$$C + C_n H_m \rightleftarrows \text{Products} \tag{9-1}$$

c) Tar, like any other carbonaceous substance, can be burned in air.

Design Criteria for Tar Cracking in Past Downdraft Gas Producers

It is noted that tar and water vapors are generated in a temperature zone of 100°C - 450°C above the air injection level in a downdraft gas producer. It is therefore extremely important (with regard to tar cracking) to generate a "hot" cross section below the pyrolysis zone so that tar and water vapors formed in the pyrolysis zone will pass through this hot layer and be partially converted. The first step towards more efficient tar cracking was therefore to incorporate a throat into the design of a downdraft gas producer. This design change is shown in Figures 9-2 and 9-3. Gumz (1950) presented a schematic of the approximate "hot" zone distribution in a downdraft gas producer (Figure 9-5) with wall tuyeres. In this context, hot temperature zone means that temperatures are at least 1000°C. These are zones in which one can expect sufficient tar cracking in a rice hull fuel bed. Figure 9-5 points out the drawbacks of this commonly used design.

a) Cold regions are generated between the tuyeres (air nozzles).

b) With this design a homogeneous hot cross section is not generated below the pyrolysis zone.

It was therefore necessary to introduce the throat as a design feature in downdraft gas producers (see Figure 9-3). Under steady-state conditions, one can expect an ideal vertical temperature distribution as shown in Figure 9-6. It is noted that the only purpose of the throat is to facilitate tar cracking. This design feature improves neither the energy content of the gas nor the kinetics of the process. On the contrary, the throat is a physical hinderance to the downward flow of the fuel. The fuel bed must be reduced in size so that they can pass through the narrow passage. The flow velocity, however, remains constant. This is the very reason why downdraft gas producers with throats cannot be used for rice hull gasification. The minimal size reduction and stable configuration of a rice hull fuel bed during thermal decomposition leads to jamming and blockage at the throat.

While writing the State-of-the-Art for small-scale gas producer engine systems for USAID the author analyzed many designs (Kaupp, Goss 1982) and came to the following conclusions:

a) It was recognized that oxygen penetration into the fuel bed is very limited. It is of the order of cm or 5 particle diameters from the point of ignition (Groeneveld 1980, Hedden 1982).

Figure 9-5. Hot temperature zone of a downdraft gas producer with wall tuyeres (Gumz, 1950).

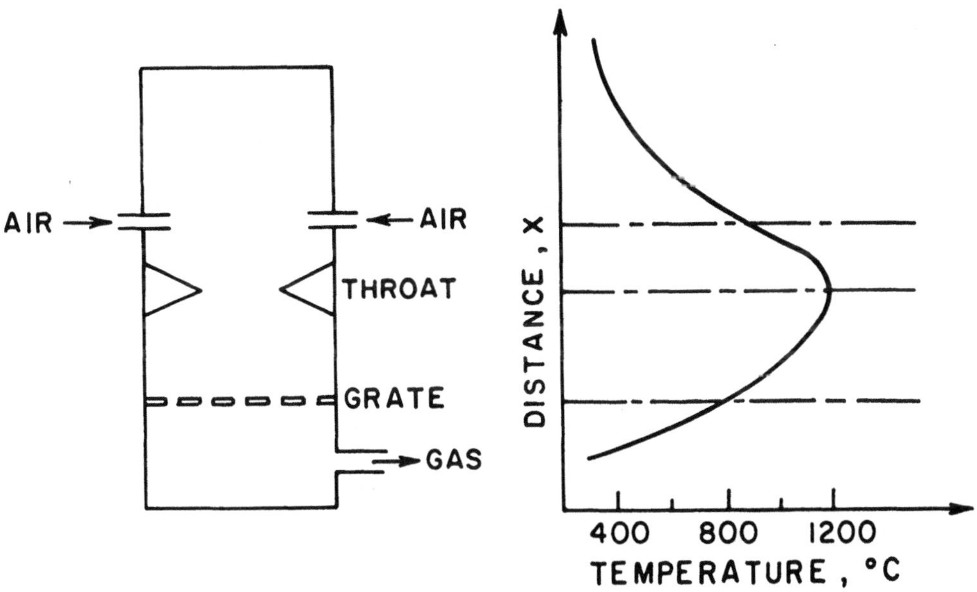

Figure 9-6. Approximate position of vertical temperature profile and throat for optimal tar conversion.

b) Groeneveld, while testing a center tuyere model shown in Figure 9-7, found a drastic increase in tar generation when the distance between the air nozzle and throat was not "right".

c) Many wall and center nozzle designs have been tried (including combinations) in order to combat the very limited oxygen penetration without much success.

d) Oxygen penetration and dispersion cannot be arbitrarily increased by increasing the air jet velocity (Groeneveld 1980).

e) The hot glowing carbon zone is mainly generated by radiation and not by combustion of carbon. It is therefore nearly impossible to scale up downdraft gas producers to large hearth diameter without losing the tar cracking potential of the unit because of cold spots.

f) The wall and middle tuyere downdraft gas producers with throats and uninsulated metal walls were a typical product of the automotive gas producer industry during the 1930's to 1950's, and were operated on charcoal. These units are not necessarily efficient in tar cracking.

g) Reliable tar conversion is dependent on the maintenance of the optimal vertical temperature profile shown in Figure 9-6.

h) It is difficult to maintain the optimum location of the "hot" fire zone as shown in Figure 9-8a (Beagle, 1980). Downward creeping fire zones, as shown in Figures 9-8b,c, are not uncommon. It is noted that the fire will always move up, independent of the design (updraft, downdraft). Downward creeping fire zones are therefore caused by uncontrollable ash removal systems.

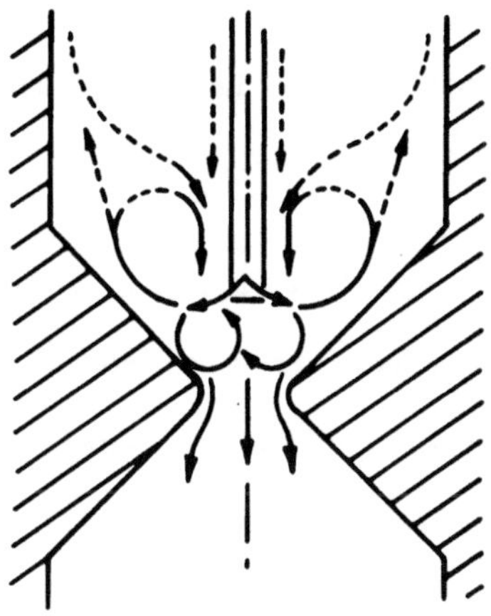

Figure 9-7. Downdraft center nozzle gas producer (Groeneveld, 1980).

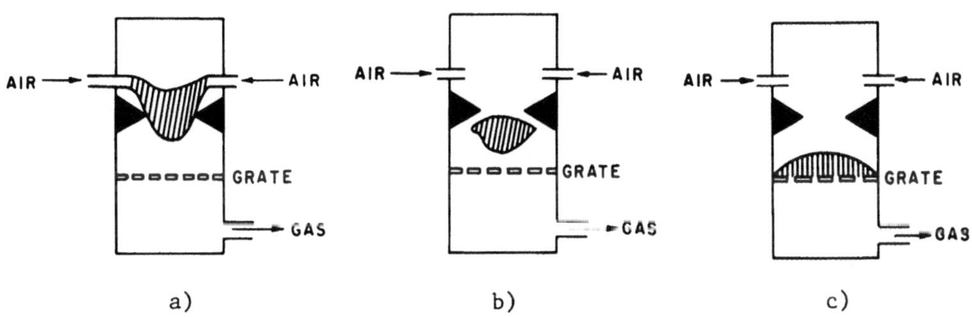

Figure 9-8. Downward creeping fire zone.

i) The danger of a fluctuating vertical temperature profile has led to designs that call for a very short distance between grate and throat. This will improve tar cracking at the expense of the reduction and shift reactions and will lead to high grate and raw gas temperatures. No reliable data is available as to the optimal distance between throat and grate.

In order to test the hypothesis that only high temperatures and a flat horizontal temperature profile across an entire cross section are needed to ensure sufficient tar cracking, special experiments were performed.

Experimental Set Up

In order to obtain an objective picture of tar cracking in a rice hull fuel bed, 38 tests were performed in the experimental set up shown in Figure 9-9. Twenty of these tests were used to draw conclusions. In order to simulate the processes taking place in a downdraft gas producer, a temperature profile similar to that observed in a downdraft gas producer was generated inside the tube.

The tube diameter was 27 mm and must be viewed as a compromise between generating a constant cross-sectional temperature profile and being able to use rice hull char and charred pellets as the fixed bed. In order to test the influence of the fuel bed structure, carbon surface and temperature on tar cracking, it was necessary to exclude the combustion of tar. A constant nitrogen flow of 0.3-0.7 liters/min was used to carry the tar-cracking products into a filter train and to exclude air.

Cases Tested

Let f be the conversion efficiency of a fuel bed where

$$f = \frac{\text{weight of recovered tar}}{\text{weight of injected tar}} \quad \text{then } f = f(\varepsilon_\beta, A_{\beta\sigma}, T). \quad \text{(See Table 9-4).}$$

The following cases have been tested:

a) $\varepsilon_\beta \approx 75\% - 85\%$, $A_{\beta\sigma} \approx 3000 - 4000\ m^2\ m^{-3}$ (charred rice hulls as a fuel bed).

b) $\varepsilon_\beta \approx 40\% - 50\%$, $A_{\beta\sigma} \approx 300 - 400\ m^2/m^{-3}$ (charred rice hull pellets as fuel bed).

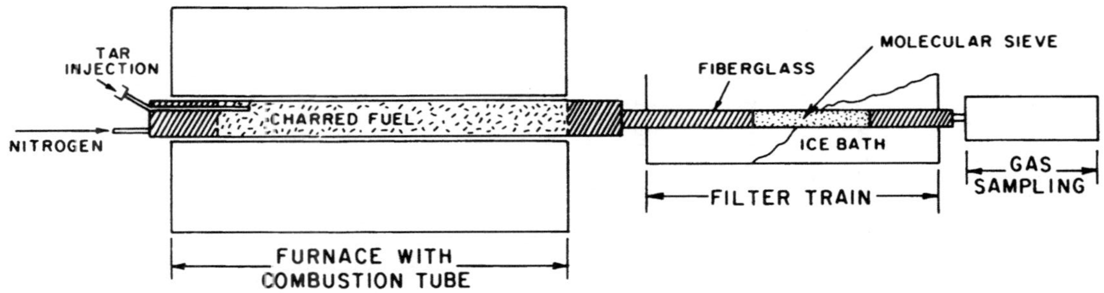

Figure 9-9. Experimental set up for tar cracking tests.

c) ε_{β} = 100%, $A_{\beta\sigma}$ = 0 (empty tube)

d) Maximum temperatures in test tube: 680°C, 800°C, 920°C for cases a) through c).

Test Material

The same type of tar was used for all tests. This tar was originally generated during the updraft gasification of rice hulls and was kept frozen until the tests were performed. Its chemical composition is given in Table 9-1.

Table 9-1. Chemical Composition of Tar Used in Experiments

Species	% weight, wet basis
Carbon C	67.86
Hydrogen H	9.31
Nitrogen N	0.85
Ash	0.20

This wet sample contained 6.18% water. It is noted that this tar is not representative of tar generated in updraft gasification. The composition of the tar, particularly the water content, is strongly dependent on the sampling location and on the temperature in the upper layers. For instance, another sampling technique was used during a different run which resulted in a tar composition as shown in Table 9-2.

Table 9-2 Chemical Composition of Tar

Species	% weight, wet basis
Carbon	17.16
Hydrogen	9.40
Nitrogen	0.44
Ash	0.70

The water content of this sample was 81.3%. The first sample was preferred for these tests because of its low water content and high viscosity.

Charred Fuel Bed

The rice hulls and rice hull pellets used as the fuel bed were precharred at 300°C and 400°C respectively in a furnace for a period of 30 minutes. An ultimate elemental analysis for the pellets charred at 300°C) is given in Table 9-3.

Table 9-3. Pellets Ultimate Elemental Analysis, 400°C.

Species	% weight, wet basis
Carbon	45.63
Hydrogen	2.51
Nitrogen	0.62
Oxygen	11.24
Residue	40.00

The pellets had a moisture content of 2.32%.

It has already been shown in Chapter 8 that the charring of rice hulls does not result in a tar-free fuel because some hydrogen and oxygen will remain in the fuel. This is again demonstrated in Table 9-3. To avoid the collection of this remaining fuel tar, special procedures were followed.

Filter Train

The wide range of tar boiling points and the remaining fuel tar make it necessary to perform the tests with a filter train that can be attached during the experiment and that can be exposed to a temperature of 400°C without losing its tar collection ability.

Based on previous experiments, a lightweight aluminum tube filled with a combination of fiberglass and 30 Å molecular sieve pellets, as shown in Figure 9-9, was both convenient and accurate.

Tar Injection

The tar was injected with an 18 gauge 3-inch spinal needle and disposable syringe. Because of the highly variable bulk density of loose rice hulls and pellets a wide range of tar to fuel ratios of 1:1.5 to 1:10 were used. It is noted that 1 kg of dry rice hulls forms about 380 grams of liquids or 100 grams of dry tar. The tar to fuel ratio in a gasifier is therefore within the tested range (see Figure 8-8). The tar was injected at a rate of 0.6 - 1 g/min.

Test Steps

a) The combustion tube was filled with about 9 gram of rice hulls or 45 gram pellets (see Table 9-4).

b) A constant nitrogen flow was established and the sample was heated.

c) When the smoke from the righthand side of the combustion tube ceased (usually at 500°C) the prepared filter train was attached. Special precautions were taken to collect tar vapors above 400°C.

d) Tar injection was initiated.

e) The nitrogen carrier gas and all gaseous products were cooled down to 7 - 12°C in an ice bath which surrounded the molecular sieve and fiberglass wool.

f) A material balance was performed. The balance used had an accuracy of one tenth milligram.

g) Samples of the gas were taken (downstream from the filter tube) during injection and analysed for CO_2, CO, H_2, CH_4, C_2H_2, C_2H_4 and C_2H_6.

Results and Discussion

Data for 20 selected tar cracking tests are given in Table 9-4. The conversion efficiency is shown graphically in Figure 9-10 as a function of temperature T with ε_β as parameter.

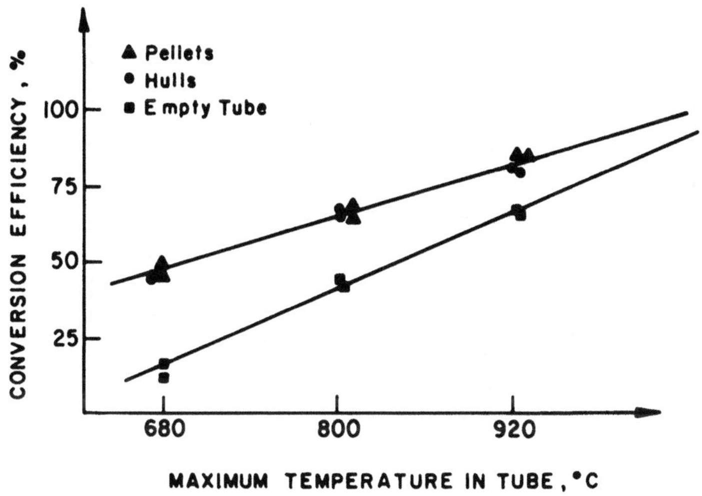

Figure 9-10. Tar conversion efficiency, experimental results.

The longitudinal temperature profiles in the tube at the three maximum temperatures of 680°C, 800°C and 920°C are shown in Figure 9-11. The odd numbers in Figure 9-11 refer to the temperature profiles in the packed tube, while even numbers refer to the profiles in the empty tube. Figure 9-10 confirms the stated hypothesis along with an additional twist.

a) Efficient tar cracking cannot be expected to occur at temperatures below 900°C. It should be noted that high temperatures

Table 9-4. Data for 20 Selected Tar Cracking Tests

Test #	Temp. (°C)	Fuel Type	Weight[1] (Gram)	Spent[2] (Gram)	*	Tar Injected (Gram)	Tar Recovered (Gram)	Conversion Efficiency (%)
1	920	Pellets	44.98	6.91	T	8.14	1.22	86.2
2	920	Pellets	27.82	4.62	T	6.85	0.97	85.9
3	920	Pellets	34.12	8.33	W	8.04	4.80	40.2
4	920	Pellets	30.03	6.68	W	8.12	6.36	21.7
5	920	Hulls	9.73	2.10	T	10.08	2.03	79.9
6	920	Hulls	8.96	2.86	T	6.42	1.12	82.5
7	920	Accuracy test with water: 5.1% Loss						
8	920	None	0	0	T	7.94	2.65	67.7
9	920	None	0	0	T	4.85	1.65	66.0
10	800	None	0	0	T	7.15	3.93	45.0
11	800	None	0	0	T	6.85	3.84	43.9
12	800	Pellets	31.95	6.20	T	7.96	2.57	67.7
13	800	Pellets	23.51	7.43	T	9.00	3.14	65.1
14	800	Hulls	8.01	1.51	T	5.92	2.01	66.0
15	800	Hulls	8.95	1.03	T	7.06	2.27	67.8
16	680	None	0	0	T	7.02	6.20	11.7
17	680	None	0	0	T	5.26	4.37	16.9
18	680	Pellets	29.61	6.63	T	3.04	1.50	50.1
19	680	Hulls	29.49	N.D.	T	6.07	3.13	48.4
20	680	Corncobs	9.99	1.43	T	6.27	2.68	57.2

*T = Tar W = Water

[1]Initial weight of the charred rice hull or rice hull pellet bed.

[2]Amount of carbon and remaining volatiles driven off during heat up period and tar cracking.

must be maintained over the whole cross section of the gas producer.

b) The availability of a carbon surface for tar cracking becomes crucial at lower temperatures but does not play a significant role at temperatures above 900°C.

c) The conversion efficiency seems to be independent of ε_β. However, it is noted that rice hulls have a much higher surface area per unit volume than pellets.

d) In the temperature range tested (680°C - 920°C) the conversion efficiency is a linear function of temperature for both the packed bed and empty tube cases.

Water Dissociation in a Hot Rice Hull Char Bed

Introduction

The moisture content of rice hulls has a marked effect on the gas quality as shown in theoretical Model I outlined in Chapter V. However, the effect of moisture content on gas quality is somewhat controversial. Jenkins (1980) in his PhD thesis argues that moisture contents of less than 20% (by weight) are necessary for fuels which are to be used in downdraft gas producers. Various other sources (Ekman 1972; Hose 1982) report that reliable operation and good gas quality were achieved in a downdraft gas producer using fuels with moisture contents of 30% - 40%.

It is noted that the term fuel moisture refers to the moisture trapped in the pores and on the surface of the fuel. This term does not include the chemically bound water which forms during the pyrolysis state. It has been shown in Chapter 8 that on the average 23-27 grams of water are released from 100 grams of rice hulls with a 0% moisture

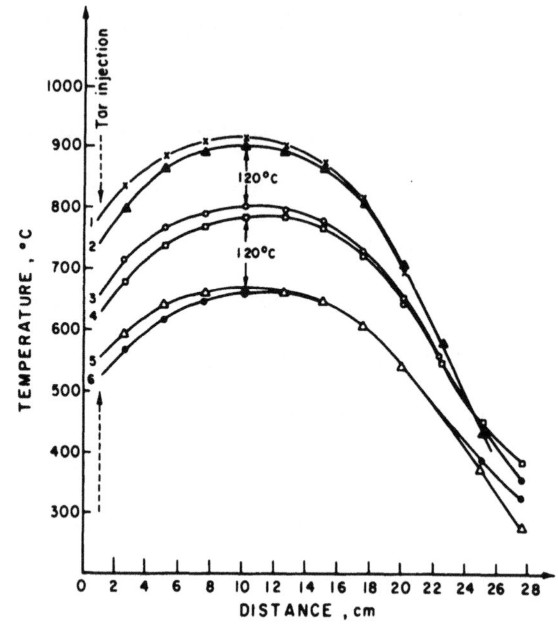

Figure 9-11. Longitudinal temperature profiles in tube.

content during thermal decomposition at temperatures of 350°C - 450°C. A qualitative assessment of the conversion of the moisture in the fuel during the gasification process results in the following conclusions:

a) The bulk of the hydrogen in the producer gas mixture comes from the conversion of water to CO and H_2 with carbon acting as a catalyst. This process may be represented by

$$C + H_2O \rightleftarrows CO + H_2 + 131 \text{ MJ/kg-mole}$$

It will be shown in the following section that considerable amounts of CO_2 and small amounts of methane are also formed. Very little hydrogen (< 1%) is formed from atomic hydrogen in the fuel, as discussed in Chapter 8. It is therefore beneficial to have a high fuel moisture content in order to increase hydrogen formation. In particular, precharred biomass fuels yield a gas which is low in hydrogen when gasified. Another characteristic of the gasification of dry fuels is the decreasing hydrogen content of the gas when a batch fed gas producer is used. These phenomena are shown in Tables 9-5 and 9-6. In Table 9-5 the gas composition from an automotive charcoal gas producer is given, while in Table 9-6 the declining hydrogen content of the gas with operating time is illustrated.

Table 9-5. Producer Gas Composition of a Wood Charcoal Automotive Gas Producer (Anonymous, 1942)

Species	% volume
CO	31.9
CO_2	1.4
CH_4	0.6
H_2	4.1
N_2	62.0

Table 9-6. Gas Composition of an Automotive Crossdraft Gas Producer. Change of Gas Composition during use (Anonymous, 1937).

	% by volume				
After a journey of	CO_2	CO	H_2	CH_4	N_2
10 kilometers	1.7	24.9	15.9	2.1	55.4
54 kilometers	2.8	26.4	11.8	2.3	56.5
198 kilometers	2.7	27.8	7.4	1.8	60.2
218 kilometers	2.4	29.1	7.1	0.8	60.8
246 kilometers	3.1	30.5	4.1	0.5	61.8
250 kilometers	2.7	29.9	3.2	0.3	63.8

b) Although the fuel moisture is the main source of hydrogen production, it has a cooling effect on the gasification process and will also lower the efficiency of the gas producer. It is noted that the water gas reaction is highly endothermic. In addition, all fuel moisture must be heated to temperatures above 700°C before significant conversion takes place, as will be shown in the following section. In a downdraft gas producer, where all moisture must pass through the hot oxidized zone (see Figure 5-1), large amounts of water may still remain unconverted. If, however, a high proportion is converted, gasification cannot be sustained due to cooling down of the fuel bed (see Table 9-6).

Example

The case of an airblown, downdraft gas producer is used in which only the following three reactions are assumed to go to completion:

$$C + O_2 + 3.76\ N_2 \rightleftharpoons CO_2 + 3.76\ N_2 - 394\ MJ/kg\text{-}mole \qquad (9\text{-}2)$$

$$CO_2 + C \rightleftharpoons 2\ CO + 173\ MJ/kg\text{-}mole \qquad (9\text{-}3)$$

$$C + H_2O \rightleftharpoons CO + H_2 + 131\ MJ/kg\text{-}mole \qquad (9\text{-}4)$$

Based on the elemental ultimate analysis of rice hulls given in Table 8-2, we can represent 30.7 kg of dry rice hulls as $CH_{1.5}\ O_{0.65}\ N_{0.01}\ (SiO_2)_{0.12}$. According to the experimental results in Chapter 8, about 27% of the initial weight of a sample of dry rice hulls is released as water during gasification. It is assumed that the rice hulls have a moisture content of 20% on a wet basis. Thus 38.375 kg of rice hulls generate 15.965 kg of water (which includes the chemically bound water). Two cases are considered: (i) 50% conversion of steam and (ii) 100% conversion according to reaction 9-4. The results are given in Table 9-7.

Table 9-7. Energy and Mass Balance

	Steam conversion efficiency	
	50%	100%
Carbon consumed for dissociation of steam (% mole)	44.3	88.7
Energy consumed MJ	58.0	116.0
Additional carbon consumed in a thermally neutral reactor (% mole)	26.2	52.5
Carbon balance (% mole)	+29.4	-41.1

The two cases show that while 50% steam dissociation is theoretically possible without combusting all the carbon, 100% steam dissociation is not possible. In addition, the example shows that the real issue in the gasification of wet biomass fuels is not their high moisture contents but is the degree to which steam conversion in the hot carbon fuel bed takes place. It is noted that it takes 4.3 MJ of energy to heat and evaporate 1 kg of water at 20°C to a final temperature of 1000°C. An additional 7.3 MJ is required to dissociate the steam to CO and H_2 by reaction 9-4. The question of steam dissociation in a rice hull char bed at various temperatures is treated in the following paragraphs.

Experimental Set up and Procedure

A slightly modified version of the experimental set up shown in Figure 9-9 was used. The modification involved the placement of heating bands at both ends and a thermocouple in the center of the tube in order to record the maximum temperature during steam injection. The test steps were:

a) The combustion tube was filled with char which had undergone low temperature combustion at about 600°C (see Table 9-7).

b) A constant nitrogen flow was maintained while the char was being heated to the appropriate temperature.

c) When the desired maximum temperature in the tube was achieved and the temperature profile (as shown in Figure 9-11) was stabilized, a known amount of water (≈ 5 grams) was injected.

d) The unconverted water was collected in a fiberglass wool-molecular sieve filter which was situated in an ice bath. The average filter temperature was 6°C - 8°C.

e) Gas samples were taken during the 700°C and 900°C test runs.

f) Steam dissociation at four temperatures (700, 800, 900 and 1000°C) was recorded.

To assure that carbon depletion did not take place during a test run, 15 grams of rice hull char were used as the fuel bed for each 5 grams of water injected. The elemental ultimate chemical analysis of the char is given in Table 9-8.

Table 9-8. Elemental Ultimate Chemical Analysis of Rice Hull Char Used for Dissociation of Steam.

Species	% weight, dry basis
C	24.56
O	3.37
H	0.21
N	0.26
Minerals	71.60

It is noted that the dissociation of 1 kg of water requires 0.67 kg of carbon according to equation 9-4. The 3.7 gram of carbon available in 15 grams of rice hull char as given in Table 9-8 are therefore sufficient and did not lead to carbon depletion during testing.

Results

While the dissociation of tar is mainly a function of temperature, as shown in the previous section, the dissociation of steam cannot be achieved by high temperature alone and requires a carbon surface. It can therefore be expected that steam dissociation will be a strong function of the fuel bed structure and carbon content of the char. No attempts have been made to examine the steam dissociation as a function of carbon availability in the fuel bed. Instead, typical char from rice hull gasification (Table 8-2) was used as a fuel bed in the test tube and the same procedures for filling the tube were applied to all reported tests. Temperature was the only parameter which was varied (from 700°C to 1000°C). Elemental ultimate chemical analysis of the char was not performed after the dissociation took place. Since steam dissociation occurred along a vertical temperature profile, typical of that found in a downdraft gas producer, it does not make sense to perform a carbon balance because of locally different carbon consumption rates. The results of the tests are shown in Figure 9-12. The squares indicate tests in which 0.5 grams of water were injected every minute, while the round dots refer to tests in which 0.2 grams of water were injected every 20 seconds. The resulting differences in steam dissociation between the two modes of injection were insignificant.

It is concluded that at low temperatures (700°C) almost no dissociation takes place (< 6%) and that large amounts of steam will be part of the raw gas leaving the gas producer. At temperatures of 1000°C, a considerable fraction (roughly 19%) of the water still remains unconverted. These experimental results are of course in complete disagreement with the results obtained from computer simulated equilibrium computations such as the one shown in Figure 9-13, where only H_2 and H_2O are specified as products.

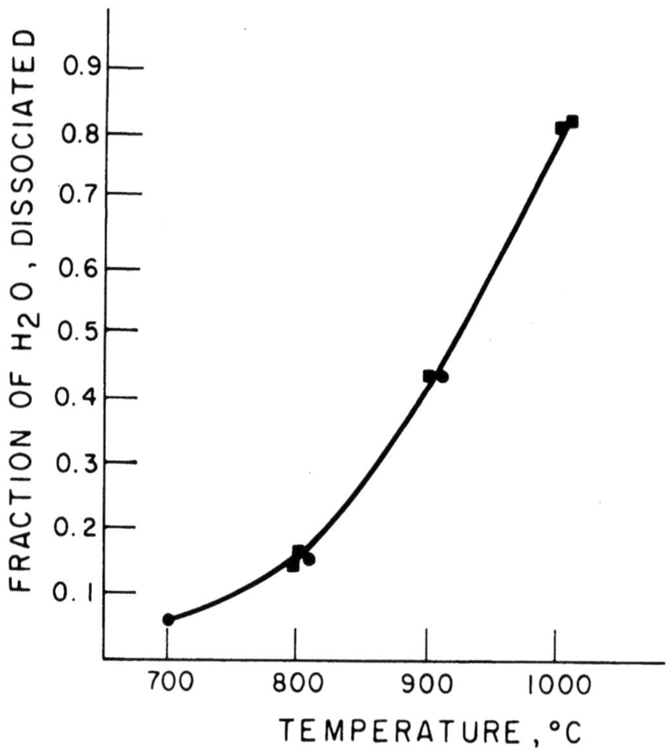

Figure 9-12. Steam conversion in a rice hull char bed.

Squares: water injection at 0.5 g every minute.
Dots: water injection at 0.2 g every 20 seconds.

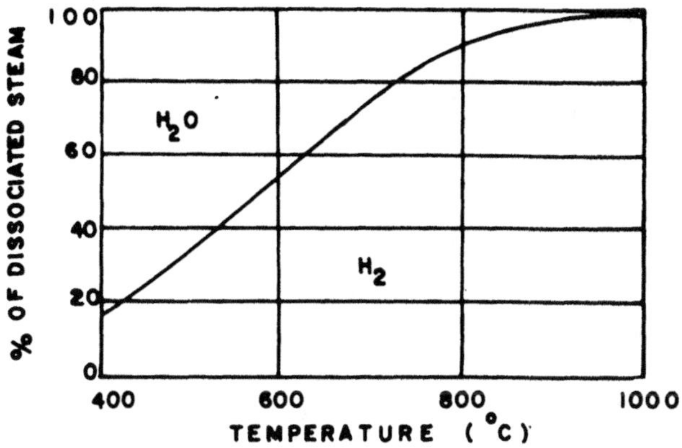

Figure 9-13. Dissociation of steam in the presence of glowing carbon (Skov, 1974).

The dependence of the equilibrium calculations on the set of products used in the computations is shown in Figures 9-14 to 9-16. The computations have been executed with the JANAF program (see Chapter 5). Three sets of products were assumed:

a) C, H_2, H_2O, CO (Figure 9-14)

b) C, H_2, H_2O, CO, CO_2 (Figure 9-15)

c) C, H_2, H_2O, CO, CO_2, CH_4 (Figure 9-16)

It is noted that the steam dissociation process does generate a considerable amount of CO_2 and CH_4 as shown in Table 9-9.

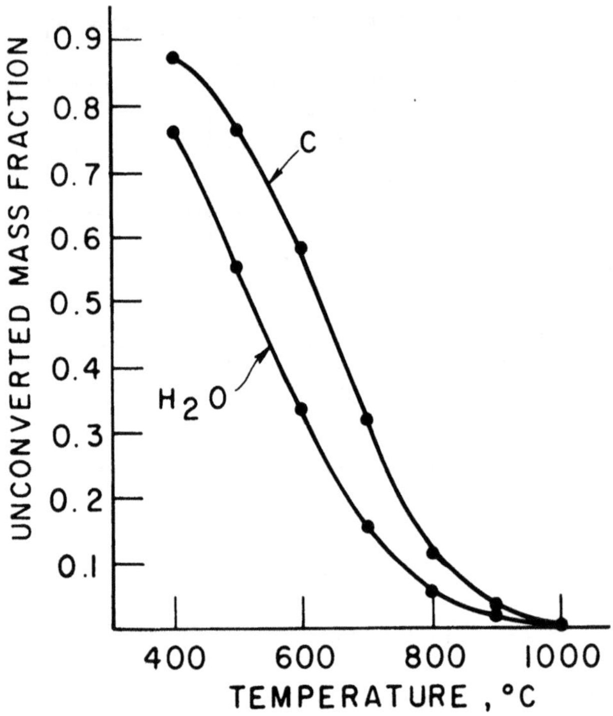

Figure 9-14. Steam dissociation with C, H_2, H_2O, CO, CO_2 as products.

Table 9-9. Gaseous Products of Dissociation of Steam at 700°C and 920°C as Analysed by Gas Chromatograph

	% volume	
Species	700°C	920°C
CO	27.1	34.5
CO_2	39.3	4.6
H_2	28.1	59.7
CH_4	4.1	1.2
CnHm	2.7	Traces

From Table 9-9 it can be seen that only option c) is a realistic set of products for the steam dissociation reactions. However, since the reaction is far from completion, the model does overestimate the converted

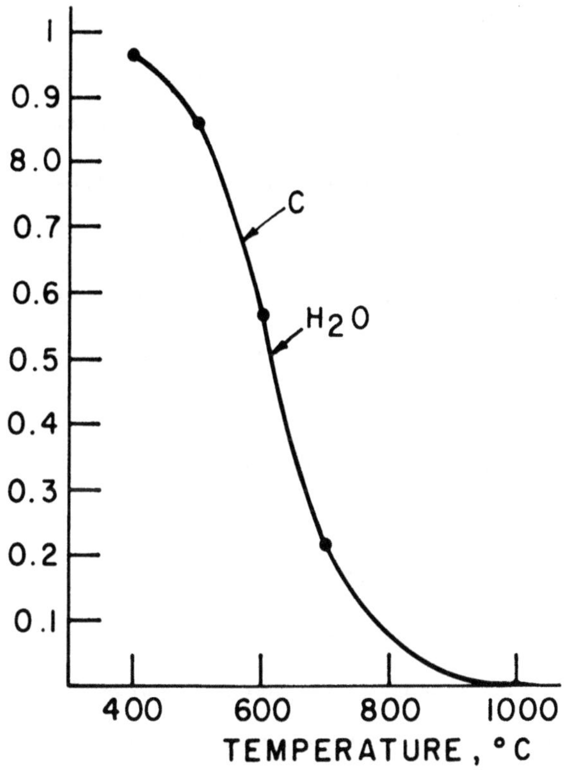

Figure 9-15. Steam dissociation with C, H_2, H_2O, CO as products.

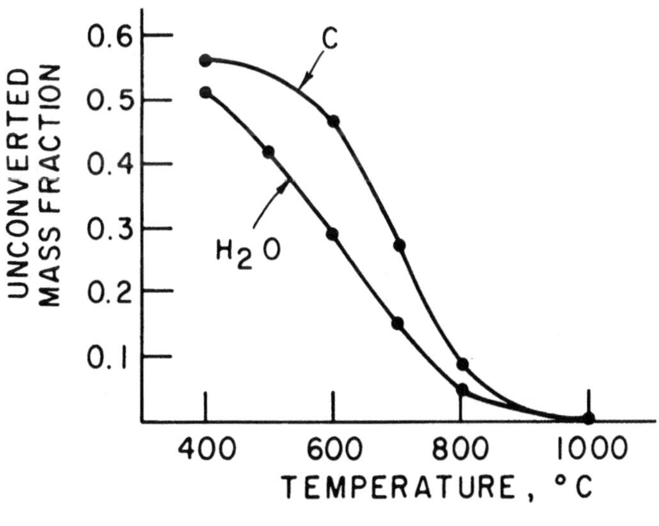

Figure 9-16. Steam dissociation with C, H_2, H_2O, CO, CO_2, CH_4 as products.

water fraction (see Figure 9-12 and 9-16). Based on these results, the water dissociation efficiency will probably be around 50% at temperatures between 900 and 950°C. The computed positive carbon balance of +29.4 (mole %) from Table 9-4 is therefore a higher bound and is not really available if one takes the following into consideration:

a) The carbon which is tied up in the condensable portion of the pyrolysis product is only accounted for if complete tar cracking takes place.

b) The gasification process does not burn off all the carbon. A fraction of the carbon remains in the char residue.

c) A fraction of the carbon is needed to make up for the energy lost by convection and radiation, the energy used to heat the fuel and air, and the energy consumed in tar cracking and pyrolysis reactions which are endothermic.

Summary

It has been argued that standard downdraft gas producers with throats will not generate a "tar-free" gas. These designs are not suitable for rice hull gasification. In addition, efficient tar conversion can only be expected at temperatures above 1000°C in a rice hull fuel bed. Tar conversion is a strong function of temperature and the high conversion efficiencies (> 95%) will not be achieved through surface reactions or catalytic effects involving the minerals in rice hulls. In addition, steam conversion does not contribute significantly to the H_2 content of the gas below temperatures of 900°C. It is therefore advisable to design a rice hull gas producer in such a way that reactor temperatures of 1000°C are achieved.

9. LIST OF SYMBOLS

Å Angstroem, 10^{-9} m

$A_{\beta\sigma}$ apparent surface area of rice hulls or rice hull pellets

T maximum temperature of horizontal temperature profile

ε_{β} void space fraction of fuel bed

f conversion efficiency of tar cracking

9. REFERENCES

1. Anonymous. 1942. Gas Producers, Automobile Engineer, November 5, 1942. pp. 433-464.

2. Anonymous. 1937. Producer gas for road vehicles. 163(4248):682-884.

3. Ekman, E. 1972. Peat as a fuel in Finland. United Nations, Economic and Social Council, Seminar on Technologies for the Utilization of Low-calorie Fuels. Varna Bulgaria. April.

4. Groeneveld, M.J. 1980. The co-current moving bed gasifier. PhD Thesis, Technische Hogeschool Twente, Netherlands.

5. Gumz, W. 1950. Gas Producer and Blast Furnaces. John Wiley and Sons. New York.

6. Hedden, K. 1982. Personal communication. Colombo, Sri Lanka. November.

7. Hose, F. 1982. Personal communication. Davis, California. August.

8. Kaupp, A. and J.R. Goss. 1981. State-of-the-Art for small-scale (to 50 kW) gas producer-engine systems. U.S. Agency for International Development, Washington, D.C.

9. Skov, N.A. and M.L. Papworth. 1975. The Pegasus Unit. Pegasus Publisher, Inc. Olympia, Washington.

10. Whitaker, S. 1972. Forced convection heat transfer correlations for flow in pipes, past flat plates, single cylinders, single spheres and for flow in packed beds and tube bundles. AICHE Journal 18(2).

10. DESIGN CONSIDERATIONS FOR A RICE HULL GAS PRODUCER

Introduction

It was mentioned in Chapter 4 that small rice hull gas producers have never been used commercially. In this context small refers to a unit with a furnace diameter smaller than 50 cm. Commercial installations like the Italian Balestra unit or the Chinese unit (which is attached to the rice mill in Suzhou in Jiangsu Province) have furnace diameters of 1.5 - 2.5 m and mechanical power outputs of 60 - 186 hp.

It is not possible to apply the large amount of knowledge acquired during the World War II experience with automotive gas producers to the design of a rice hull gas producer. The differences between the feed stocks and the types of applications are striking and are pointed out here again. A typical World War II downdraft gas producer had the following operational characteristics:

a) Good fuels: charcoal, hardwood, coke, and anthracite were used.

b) A gas output of 200 $Nm^3 h^{-1}$ was not uncommon in 40 cm diameter furnaces. This means that up to a 100 hp engine could be driven with this unit.

c) World War II units operating on wood, charcoal, coke, or anthracite did not have char removal problems. The residue generated by the gasification process amounted to only 1% - 5% (by weight) of the feed input. This small amount of residue could either be shaken down through a grate (as shown in Figures 10-1 through 10-3) or simply left at the bottom of the gas producer and removed periodically (2-5 hours). The latter

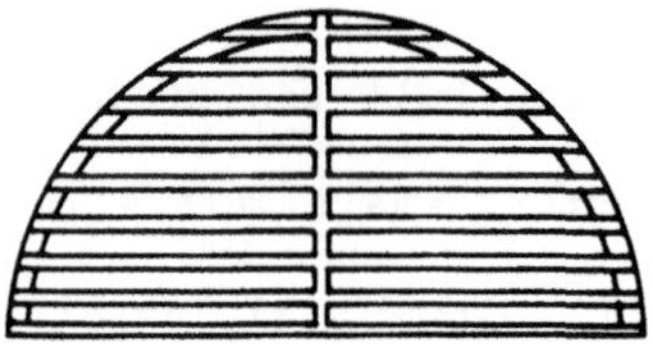

Figure 10-1. Flat slot grate (Rambush, 1923).

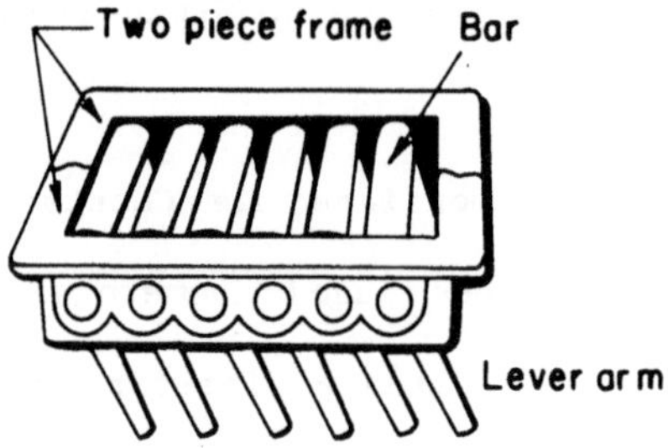

Figure 10-2. Shaker grate (Skov, 1975).

Figure 10-3. Shaker grate (Schläpfer, 1937).

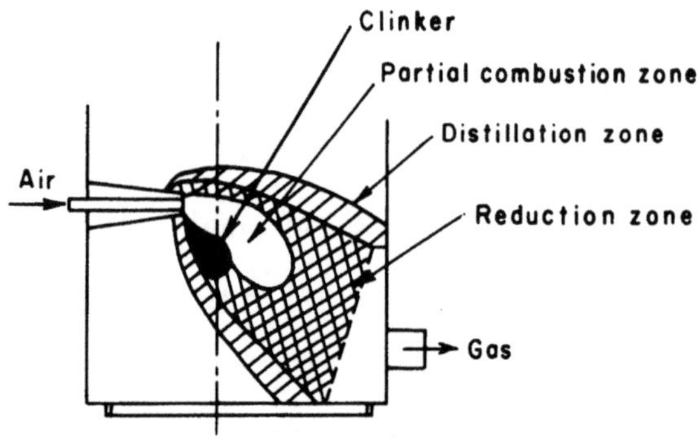

Figure 10-4. Crossdraft gas producer with no grate (Harter-Seberick, 1937).

option was the most common procedure for residue removal in crossdraft gas producers, as shown in Figure 10-4.

d) The high bulk density of the fuel (≈ 250 - 300 kg m^{-3} for wood, 500 - 800 kg m^{-3} for coal) enabled one to design units with small fuel hoppers that could operate for 2-4 hours without refueling.

e) A small firebox and an appropriately located throat in a downdraft gas producer permit the development of a homogeneous fire zone which aids the partial cracking of tar and dissociation of water (Chapter 9).

On the other hand a typical rice hull gas producer has the following operational characteristics:

a) A fuel is used which has a high natural ash content (15% - 23% by weight) and a low bulk density (100 kg m^{-3}).

b) A gas output of 200 Nm^3 h^{-1} is achieved in large diameter furnaces (> 1.5 m).

c) The residue after gasification amounts to 30% - 40% of the initial volume of the feedstock and 25% - 35% of its initial weight. An effective continuous ash removal system is therefore required.

d) The low bulk density of rice hulls necessitates either a very large hopper or a mechanism for continuous feeding.

From previous chapters it is concluded that the three parameters which influence rice hull gasification the most are the following:

a) Continuous ash removal.

b) Superficial gas velocity.

c) Diameter of the gas producer.

This work has concentrated on very small-scale units, having furnace diameters of 15 - 30 cm. Nevertheless it is useful to describe in detail two large rice hull gas producers that have been operating commercially for quite some time and for which some performance data exists.

Italian Balestra Type Updraft Rice Hull Gas Producer (1910 - 1944)

Although this unit has been discussed in several papers, two papers, one by Garbarino (1911) and the other by Borasio (1944), have been translated to extract relevant information. A schematic of this rice hull gas producer-engine system is shown in Figure 10-5. The feeding mechanism consisted of an elevator (b) which transported the rice hulls into the small hopper on top of the gas producer. Large amounts of air leaking through the feeding arrangement could easily cause an explosion in the gas producer. Thus it was essential to have the feeding column always filled with rice hulls. The large cylinder reaching into the gas producer was necessary in order to keep the rice hull bed always at the same level. Raising the rice hull fuel bed would restrict the disengagement zone and would result in a large carryover of fine particles (see Section 10.4). Lowering the level could lead to the development of a channel through the fire zone which would in turn cause poor CO_2 reduction. The knife gate (c) was kept open during operation. The diameter of the furnace was 1.5 m at the bottom, 1.9 m at the top. It had a refractory lining as shown in Figure 10-5. The ash removal system consisted of two spiral-shaped wipers (g and h) made out of cast iron. The wipers rotated in a continuous step-like motion generated by a belt

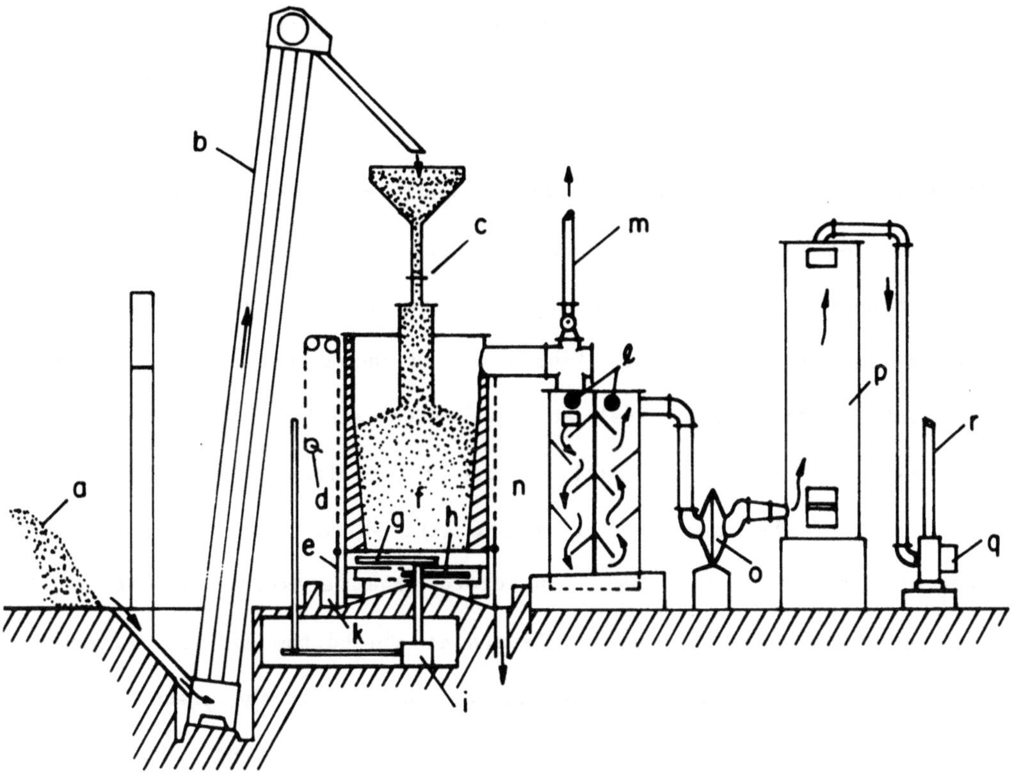

Figure 10-5. Italian updraft rice hull gas producer of the Balestra type (Garbarino, 1911).

drive using transmission i. This motion caused partial crushing of the charred rice hulls which could then fall through. The rice hull ash then slid down the cone-shaped bottom of the gas producer and was flushed out with water in the ring channel (k). It is noted that this gas producer did not have a water seal at the bottom, since it was an updraft suction gas producer. One interesting feature of this unit is the way in which the air was introduced at the bottom. The unit had a cylindrical steel sleeve (e) at the bottom which could be raised and lowered by the drive (d). Full opening of the sleeve exposed the entire lower part of the gas producer to the air, while closing of the sleeve prevented air from entering. Therefore, the necessary combustion air had to move through the two ash layers at the bottom and was introduced into the center through the step-like arrangement of the grate.

After leaving the gasifier, the producer gas entered a cooling and expansion tower (n). It is noted that this tower's main function was to cool the gas from an inlet temperature of 130°C to 40°C. The arrangement consisted of a series of baffles with two water spray nozzles at the top (l). The spray caused condensation of most of the tar which was then collected in the centrifugal tar separator (o). Two types of gas cleaning trains (p) were used: a dry packed bed (coke) and a wet packed bed (coke, with spray nozzle on top). Because of the sizeable pressure drop across the system, a suction fan (q) was sometimes installed at the end of the gas cleaning train. This procedure increases the volumetric efficiency of the engine.

It is not clear from the literature what the actual horsepower output of this gas producer was. Horsepower outputs of 25 - 60 hp are mentioned. An average gas output of 160 m^3 h^{-1} is reported. However, no reference to the temperature of the gas is given. In Chapter 1 it was noted that there are many ways to operate a gas producer and that a reliable way of describing the system is by the equivalence ratio ϕ. The combination of an updraft gas producer and a gas engine requires extensive gas cleaning because of the large amounts of tarry liquids generated which must be removed from the gas stream. However, one should not conclude that the reason for building an updraft unit was because downdraft gas producers did not exist. The decision to build an updraft unit was also based on the fact that a highly valuable by-product, tar, was generated. It is noted that the gasification of 100 kg of rice hulls yielded 122 - 158 kg of producer gas, 2 - 4 kg of "Catrame" and 20 - 25 kg of char. It is worthwhile to point out that Catrame is the Italian word for the material which is commonly used for road surfaces (asphalt). During the 1910 - 1950's Catrame was a very valuable product. It was used to pave barn floors and small backyards, and to do repair work on roads. Thus, the operators of Italian rice hull gas producers did not wish to suppress tar generation. In fact the units were operated at very low furnace temperatures (700 - 900°C) which yielded a fair amount of tar. The composition of the producer gas generated as well as its heating value is given in Table 10-1.

Table 10-1. Producer Gas Composition

Species	% Volume
CO	25.0 - 26.1
CO_2	4.2 - 4.5
H_2	4.8 - 8.6
CH_4	7.1 - 7.2
Cn Hm	Traces
N_2	51.2 - 56.1
O_2	2.1 - 2.5
HHV	6.25 MJ kg^{-1} (6.72 MJ m^{-3})

It is noted that the gas has a very high heating value and contains a considerable amount of methane. Based on the given mass balance, it is concluded that the unit operated at a ϕ of about 0.3. Considering all this one can conclude the following:

a) The unit was operated such that the superficial gas velocity through the bed was very low at roughly 5 cm s^{-1} - 10 cm s^{-1} (see Chapter 6).

b) Furnace combustion temperatures were low (in the 700 - 900°C range) which resulted in a producer gas with a high heating value of 5.4 MJ kg^{-1} to 6.25 MJ kg^{-1}.

c) Tar was considered a valuable by-product and not a pollutant or health hazard.

d) The gas cleaning train and centrifugal tar separator were capable of cleaning the gas to the extent necessary for use in an internal combustion engine.

e) The gas engine powered all equipment in the rice mill. Not all the gas which was generated could be used and some was simply flared off through pipe (m) or used for heating service (Figure 10-5).

Rice Hull Gasification Plant in Suzhou, Jiangsu Province, China

This plant is described in detail in the FAO report RAPA No. 61, Bangkok, Thailand, June 1982. A schematic diagram of its design is given in Figure 10-6. It is noted that this unit operates in the downdraft mode with the air inlet (1) on top. Because of the large quantities of ash which must be removed from the unit, it was necessary to set the entire plant in a water jacket, which creates a hydraulic seal. The char gradually settles to the bottom of the water jacket and is then pumped up as a slurry (8) with a common mortar pump.

The grate design is different from the Balestra unit. Instead of having one or two wipers rotating, the sieve plate in the Chinese unit itself rotates. This grate also features an annulus through which the gas is drawn. The annulus also provides a disengagement zone for larger particles in the gas stream. This method of drawing off the gas has been widely used in automotive downdraft gas producers. Despite the water jacket, the gas exit temperature is still high at 300 - 400°C. The recorded gas composition is given in Table 10-2.

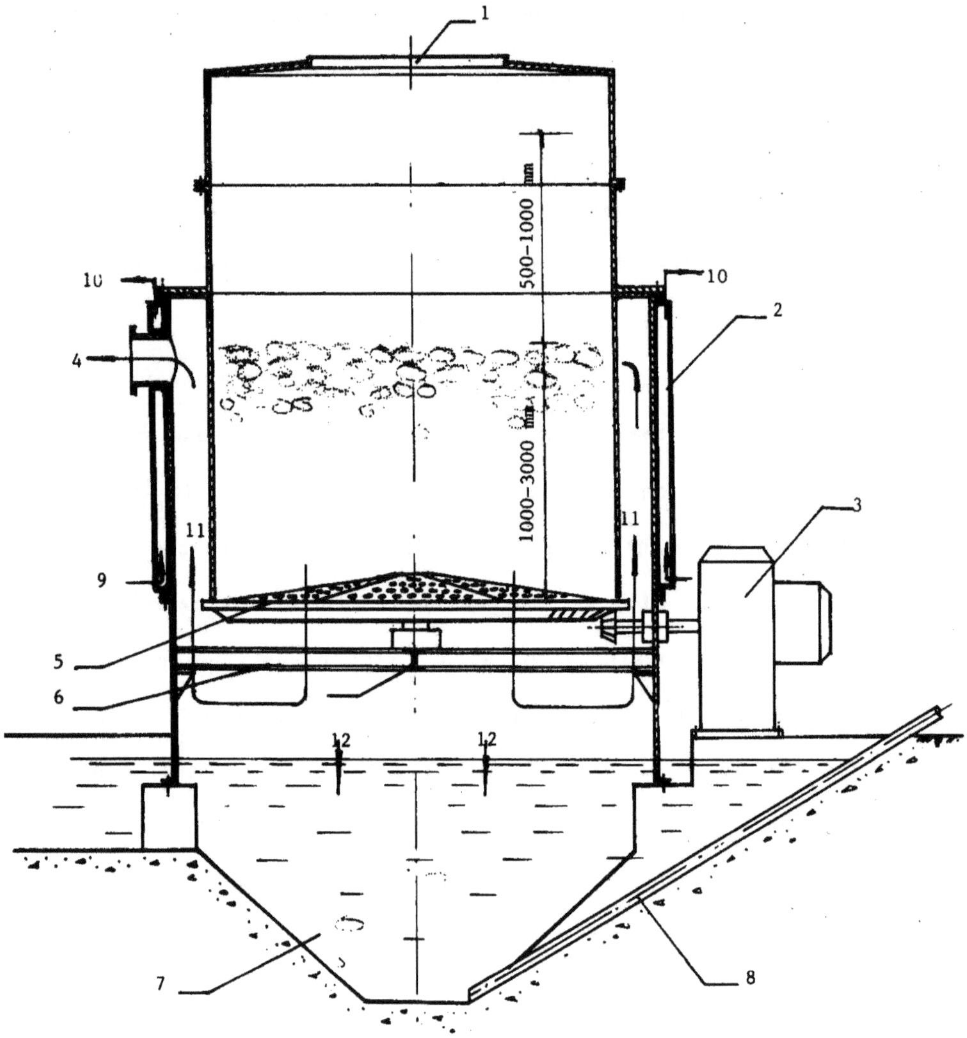

1. Fuel & Air inlet
2. Cooling water jacket
3. Furnace grate reduction box
4. Gas outlet
5. Rotary grate
6. Grate support
7. Ash settling pond
8. Ash removing tube
9. Cooling water intake
10. Cooling water effluent
11. Gas
12. Ash

Figure 10-6. Chinese rice hull gas producer (Rapa No. 61).

Table 10-2. Composition of Producer Gas From Downdraft Chinese Rice Hull Gas Producers

Species	% Volume
CO	15.4
H_2	9.5
CH_4	0.4
O_2	2.5
CO_2	7.25
N_2	65
LHV (lower heating value)	2.6 MJ kg^{-1} (3.0 MJ m^{-3})

It can be seen that the gas is of very poor quality with respect to its heating value and has only about one-half the heating value of the gas from the Balestra unit. This rice hull gas producer can be viewed as an updraft gas producer turned upside down. The raw gas from this unit should contain less tar than the gas from updraft units. It is claimed that the extensive filtering system will reduce the impurity content of the raw gas to less than 30 mg Nm^{-3}. Due to the low heating value of the gas, the gas-engine generator set consumes 5.1 m^3 of producer gas per Kilowatt-hour. However, it is noted that the efficiency of the overall plant is good at 13% (61% gas producer efficiency, 22% engine efficiency computed from available data). The underlying reason for installing this plant was two-fold:

a) A rice mill with a capacity on the order of 50 ton/day paddy powered by a rice hull gas producer like the one described above saves 3300 - 4180 barrels of diesel oil per year.

b) The same rice mill produces 3571 tons of hulls a year. Therefore, 0.5 - 0.7 hectares of land would be needed to store these rice hulls, which cannot be burned in the open. The gasification process reduces the rice hull volume by 60%.

Most rice mills in the Developing Countries are much smaller than the ones mentioned above (1 - 2 ton/day). It is therefore useful to determine the design criteria for a small rice hull gas producer that could power a 5 - 15 hp engine. In order to optimize the gas producer-engine system, it is necessary to treat the various parts of a gasifier as separate components which do not interact with one another.

The author has been testing various gas producers, gas cleaning trains and ash removal systems and has explored the following:

a) Continuous ash removal systems.

b) Ash removal by vibration.

c) Force feeding systems.

d) Downdraft gas producers without throat.

e) Various air injection systems.

f) Various gas cleaning trains.

g) Gas exit designs for rice hull gas producers.

Design Considerations for Ash Removal Systems

Introduction

Because of the large amounts of rice hull char that must be removed during the gasification process, a continuous ash removal system is essential. Experiments with an open core 15 cm diameter gas producer showed that the residue retained 67% of the initial volume of the feed input, if the caked rice hulls were not disturbed. If the fuel bed was stirred, the residues retained 37% of their original volume, i.e. before gasification. After the char was discharged into the ash bin, it occupied 32% of its original volume and had a bulk density of 117 kg m^{-3} (dry basis). The volume fraction occupied by the char is denoted here by E. It is now of interest to compute the velocity v_r of the charred residue through the rice hull gas producer as a function of the rice hull consumption rate R (kg h^{-1} as received), the gas producer diameter d, and the volume fraction E. We have therefore:

$$v_r = \frac{1.27\ E \cdot R}{d^2 \cdot \rho_b}$$

where ρ_b = bulk density of the rice hulls as received.

Examples

A 15 cm diameter gas producer operating an engine with an output of 5 hp requires about 10 kg of rice hulls per hour (ρ = 120 kg m^{-3}). This would result in:

$$v_r = \frac{(1.27)(0.37)(10}{(0.15)^2\ (120)} = 1.74\ m\ h^{-1}$$

On the other hand, the large Italian rice hull gas producer with a fire-box diameter of 2 m and a power output of 180 hp would require 540 kg h^{-1} of rice hulls (Beagle, 1978). This results in:

$$v_r = \frac{(1.27)(0.37)(540)}{(4)(120)} = 0.53\ m\ h^{-1}$$

From the two examples it is obvious that the ash removal velocity is significantly higher in smaller units. It is noted that the two ash removal systems discussed in Sections 10.1 and 10.2 have limitations with respect to the rate of char removal.

Eight different types of rice hull char removal systems have been tested and the three most successful are discussed here.

Ash Removal Designs

Parameters which influence the grate design are shown in the diagram in Figure 10-7. In general, the following ash removal options are available:

a) No grate at all.

b) Fixed grate allowing for periodic removal of the ash.

c) Mechanical, rotating grate allowing for periodic removal of the ash.

d) Mechanical, rotating grate allowing for continuous ash removal.

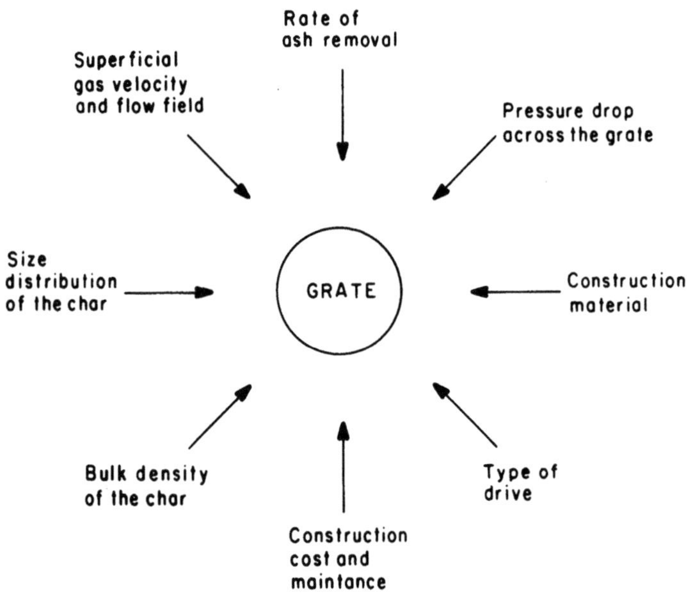

Figure 10-7. Parameters which influence the grate design.

Option a), which has been widely used for fuels with very low ash contents, is impractical for continuous rice hull gasification because of the large volumes of char generated during gasification. Options b) and c) also result in unsatisfactory gas producer performance because of the high char velocities of 0.4 - 1.8 m h^{-1}, as previously outlined. It is noted that based on the above calculations, periodic ash removal at 30 minute intervals would cause a fire zone rise of 0.20 and 0.9 m for the range of char velocities and lead to a breakthrough of the fire zone at the top which in turn would result in deteriorating gas quality. The gasification process will therefore be sustained only for a short period of time in rice hull gas producers with fixed grates or with periodic ash removal systems.

Based on these observations, it was concluded that a controlled continuous ash removal system is necessary for reliable continuous gasification of rice hulls.

When dealing with a difficult fuel such as rice hulls, it is very important to eliminate factors that may influence the performance of a grate, but are not caused by the design of the grate. For instance, even a well-designed grate will not help if the rice hull fuel bed above the grate cakes. In this case, gravity flow stops altogether. In order to avoid misinterpretation of grate performance, the grates were first tested on a cold test stand that simulated a 15 cm diameter downdraft gas producer (Figure 10-8). A good ash removal system should meet the following criteria:

a) It must remove the char in any form (loose, lumped, mixed with small pieces of slag.
b) The rate of ash removal must be controllable.
c) Because force feeding of rice hulls into the firebox column is

Figure 10-8. Cold test stand for grate performance.

one option with regard to overcoming gravity flow problems and channel formation, the ash removal system should be able to remove densified char.

d) The design should be simple.

Design I

The design shown in Figure 10-9 is a copy of the large grates used in past coal gasifiers. It was used in a 30 cm diameter downdraft gas producer of the type showed in Figure 5-1 by Williams and Horsefield (1982) to test the gasification characteristics of California crop residues. The version shown in Figure 10-9 fits a 15 cm diameter gas producer and can be installed as shown in Figures 10-10 and 10-11.

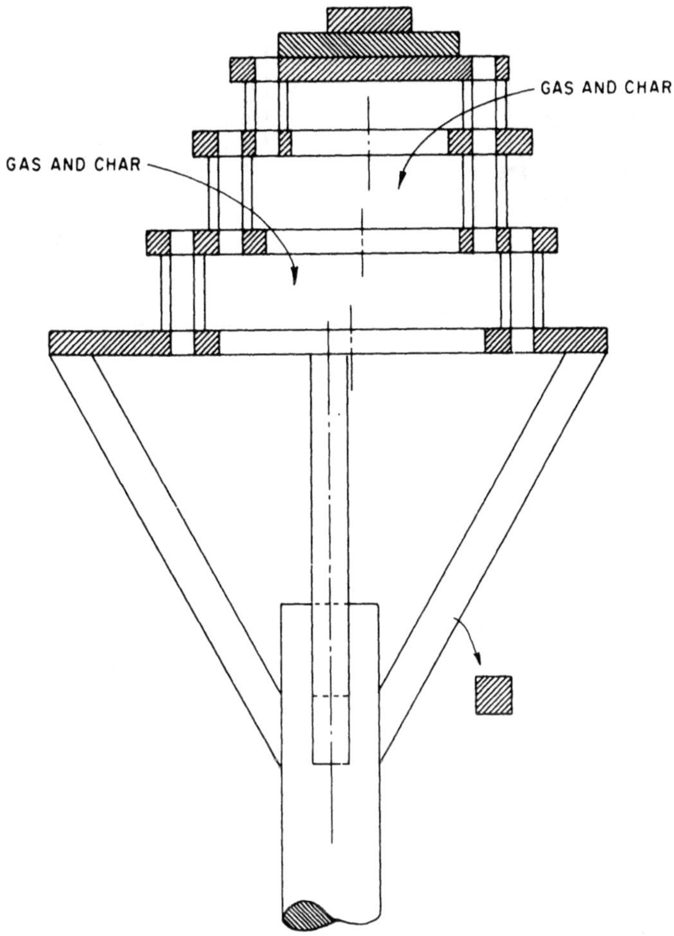

Figure 10-9. Eccentric rotating grate.

1. Performance Characteristics and Properties.

The eccentricity of the rotating grate causes the char to be pressed through the horizontal slats. The construction of this fairly complex grate takes about 20% of the total time needed to build a small 15 cm diameter gas producer. Although the grate is very effective for wood chips and corncob char, it handles rice hull char in a very erratic

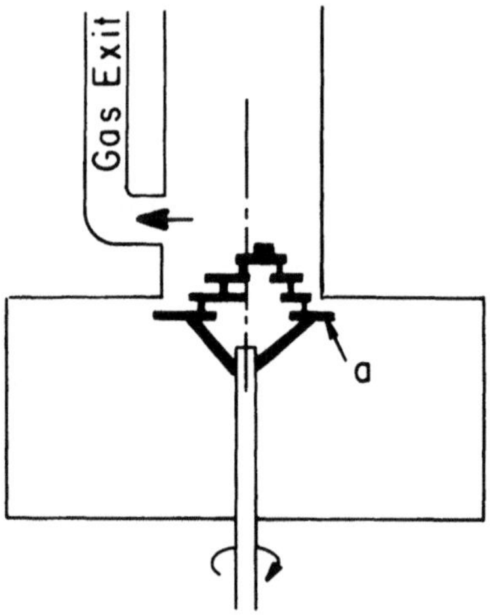

Figure 10-10. Gas exit above grate.

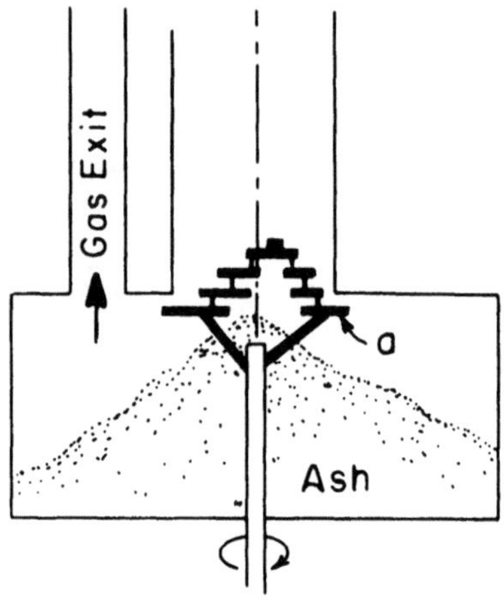

Figure 10-11. Gas exit below grate.

manner. One of the major difficulties is that it cannot remove consolidated matter. In such a case the grate generates free caves or simply plugs up. Because the superficial velocities between the plates are high in a small gas producer, it is not advisable to pass the gas through the grate as shown in Figure 10-11. One big advantage of the eccentric grate is that it slowly and continuously stirs the lower part of the gas producer. It is therefore suited for gas producers that operate at low temperatures (700 - 800°C) and lower ϕ's. The author considers this design a non-optimal solution to the problem.

Design II

A completely different approach to the problem is represented in Figure 10-12. This design consists of a solid disk which is fixed at a distance d below the bottom of the firebox. A curved wiper slowly rotates over the top of this disk. It is noted that the ash is not removed through holes in the disk but is pushed over the edge of the disk and into the ash bin.

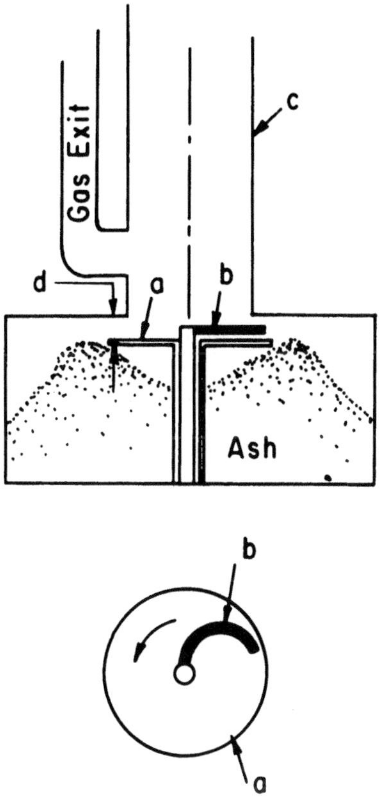

Figure 10-12. Grate with rotating curved wiper.

1. Performance Characteristics.

This grate was tested extensively on the cold test stand since it holds the most promise with regard to meeting the criteria mentioned above. The results are given in Figure 10-13. Rice hull char was used for these tests. The wiper height was 7 mm and the disk diameter was 30 cm. The firebox column had a diameter of 15 cm. This grate design removes the char in a very steady, continuous manner. It has the following advantages:

a) The amount of char removed can be controlled by four parameters:

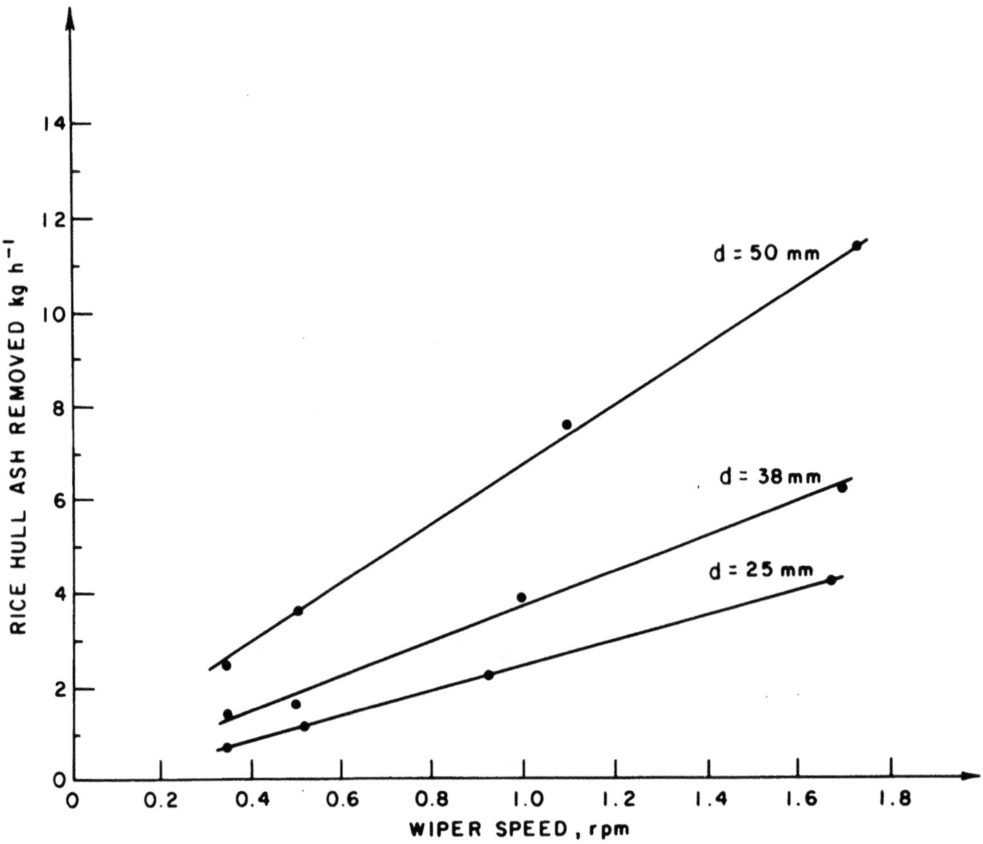

Figure 10-13. Rice hull char removal as a function of the wiper speed and distance [d]. Wiper height 7 mm, disk diameter 30 cm and firebox diameter 15 cm.

i) The overlap between firebox diameter and disk diameter.

ii) The distance d between the firebox and the disk.

iii) The height of the wiper.

iv) The wiper speed.

b) This system is capable of removing consolidated matter. The wiper simply slices it off piece by piece. It can also handle slag as long as the slag is not larger than the distance d.

c) It makes possible removal of char which has been compacted by force feeding of the rice hulls into the firebox.

d) The system is simple to build and easy to replace.

The main disadvantage of this type of grate is the ease with which the fuel bed can be disturbed near the grate. Because of the high pressure drop caused by passing the gas through the space between the disk and the bottom of the firebox, it is recommended that the gas be drawn off well above the grate, as shown in Figure 10-14.

Design III

This design is a slightly modified version of Design II. It is shown in Figure 10-14. This design allows better utilization of the ash bin. It slices the lowest layer of the char off the plate and is therefore also suitable for periodic ash removal in low ash fuel gasification.

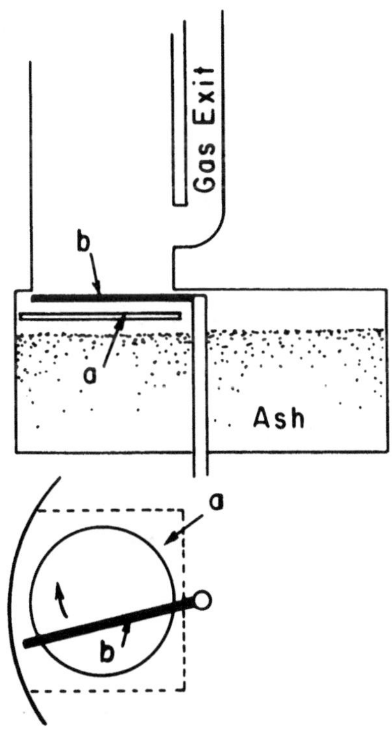

Figure 10-14. Eccentric wiper grate.

Design IV

Small gas producer-engine systems (5 - 15 hp) cannot be justified by their fuel savings alone, they must be very simple and very cheap. It is therefore important to explore the possibilities for removing the ash with a grate that does not need an external drive. In this small range, the engine and gas producer can be mounted together on a free-swinging frame. Here, the engine vibration facilitates ash removal. It is noted that past automotive gas producers had very few difficulties with regard to fuel flow and bridging of the fuel in the fuel hopper. This was due to engine vibration and perhaps to rough roads. In order to test the effect of engine vibration on the gasification process and on ash removal, the engine and gas producer were mounted together on a free-swinging frame as shown in Figures 10-15 and 10-16. It is important to leave the bars unbolted at the ends because this would inhibit vibration in the gas producer. The small Briggs and Stratton 10 hp engine caused a very satisfactory vibration in the corn-cob fuel bed and also in the grate disk. The ash removal over the disk can be fine tuned by varying the distances d and a. It is noted that the gas should be drawn off above the grate because even slight adjustments of the distance d resulted in large pressure drop changes across the grate.

Another version of the vibrating grate is shown in Figure 10-17. This version is a slightly bent disk that facilitates the char removal by vibration better than the flat disk in the case where the distance d (Figure 10-12) is very small.

The results of two tests with rice hull char and corncob char are given in Table 10-3.

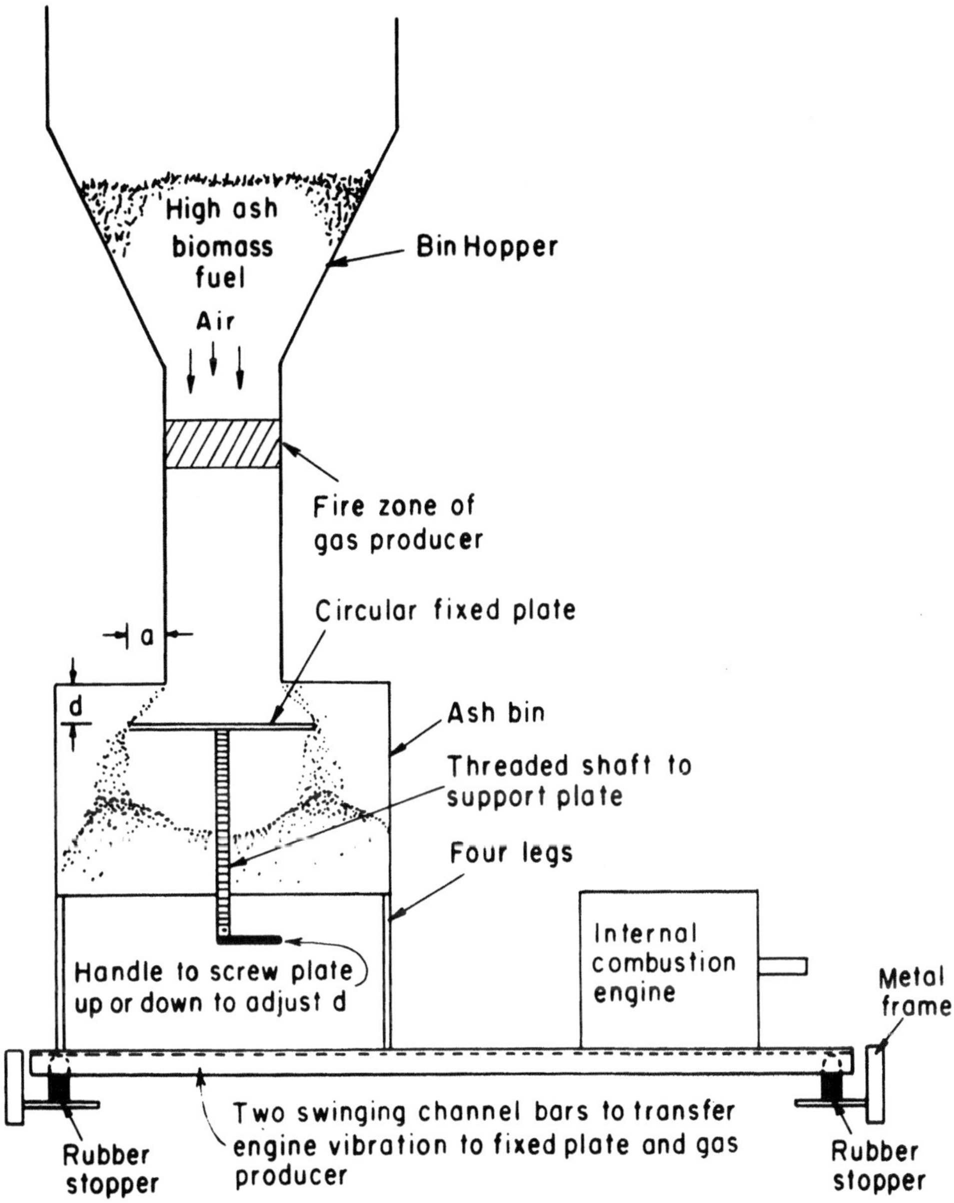

Figure 10-15. Gas producer and engine mounted on a free-swinging frame.

Figure 10-16. Detail of free-swinging bars and mounted engine.

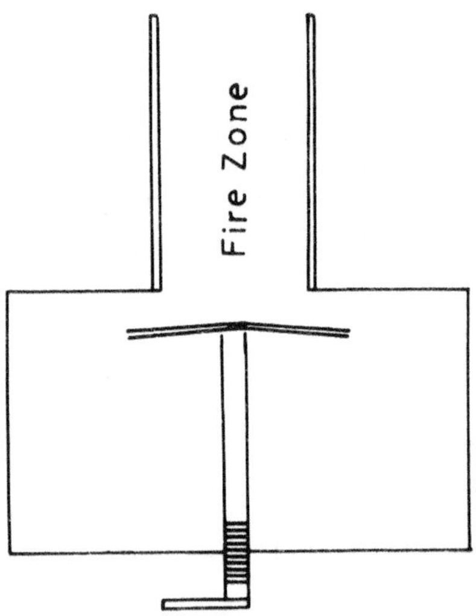

Figure 10-17. Slightly curved vibrating disk.

Table 10-3. Char Removal by Vibration

Fuel	Char Velocity $m\ h^{-1}$	Mass Flow Rate $kg\ h^{-1}$
Corncob char	0.71	2.0
Rice hull char	1.88	7.4

Distance, d, was set at 38 mm. Distance, a, was 63 mm for this disk. The tests showed the potential for simply using the engine vibration to induce horizontal char removal by means of an excited disk underneath the firebox. This method of char removal will be specific to the particular gas producer, to the engine, and also to the way in which the two are mounted on a free-swinging frame.

Summary

Of the eight grates tested, four have proven reliable. It is noted that Design I stirs the lower part of the char bed whereas the other three designs have no mixing effect on the bed. However, Design I is also the most unreliable with respect to a constant ash removal rate. In small units the gas should be drawn off above the grate with these designs. Passing the gas through the grate will result in high superficial gas velocities that cause entrainment of char particles. These char particles are difficult to separate from the gas stream.

Design Considerations for the Gas Exit

One of the most important obstacles to the use of producer gas in an internal combustion engine is the impurities in the gas. In general, it is not difficult to generate a combustible gas. However the large amounts of water vapor, tar, char particles and solid carbon entrained in the gas stream cause serious operational difficulties in the gas cleaning apparatus and in the engine. This section addresses exclusively the topic of separation of fine particles from a gas stream by settling. It is noted that past gas producers of the updraft design had a large disengagement zone as shown in Figure 10-5 for the Italian Balestra unit.

Downdraft units such as the Chinese rice hull gas producer and the automotive Imbert gasifier use an annulus to disengage particles and to act as a heat exchanger for the sensible heat of the gas (Figure 10-6). On the other hand, a common complaint about past crossdraft gas producers had to do with the excessive amounts of solid particles entrained in the gas stream.

The question as to whether a low superficial exit gas velocity and a long vertical disengagement zone facilitate separation of char and gas can be answered quite accurately on a theoretical basis. The usual simplifying assumptions that allow an exact representation of the terminal velocity of a particle are:

a) The particle is spherical and rigid and no slip exists between the particle and the gas.

b) The gas is homogeneous within the length scale of the particles.

c) The settling velocity is so low that all inertia effects are negligible.

d) The flow field is laminar.

Of course, it is obvious that one of the assumptions, a) is not at all satisfied for rice hull char. But as pointed out by Heiss (1952) even for nonspherical particles where the terminal velocity depends on the orientation of the particle, the highest ratio between maximum and minimum velocity is less than 2.

Example

For a typical gas exit temperature of 600°C the average viscosity of the producer gas will be $\eta = 356° \; 10^{-7}$ kg m^{-1} s^{-1}, as shown in Figure 10-18. Assuming that the gas behaves according to the ideal gas law, the calculated gas density ρ_g is 0.33 kg m^{-3}. The terminal velocity of a particle may be computed using Stokes theorem if Re < 0.2. We therefore need to look at particles with diameters

$$d < \left(\frac{3.6 \; \eta^2}{(\rho_s - \rho_g) \; \rho_g g} \right)^{\frac{1}{3}}$$

The density of rice hull char, ρ_s, has been determined to be between 2.0 - 2.3 g cm^{-3} depending on the amount of carbon left in the char (2.3 g cm^{-3} is the true density of the rice hull ash). Consequently d < 85 µm for laminar flow settling. The terminal velocity of a 85 µm particle is given as

$$u_t = \frac{(\rho_s - \rho_g) \; g \; d^2}{18\eta} = 0.25 \text{ m s}^{-1} \qquad (10\text{-}1)$$

If we assume that the particle accelerates from u = 0 to u_t within the gas stream we have:

$$u(t) = \frac{(\rho_s - \rho_g)\, g\, \rho_s\, d^2}{18\eta} \left(1 - e^{-\frac{18\eta}{\rho_s\, d^2}\, t}\right) \qquad (10\text{-}2)$$

this equation yields:

$$t^* = \frac{4.6\, d^2\, \rho_s}{18\eta} \qquad (10\text{-}3)$$

where t^* is the time needed for the particle to reach 99% of its terminal velocity. In this case:

$$t^* = \frac{(4.6)(85)^2(10^{-6})^2(2300)}{(18)(356)(10^{-7})} = 0.1\ s$$

The distance traveled when the particle reaches 99% of its terminal velocity is:

$$h = 0.785\ u_t \cdot t^* \qquad (10\text{-}4)$$

$$= (0.785)(0.25)(0.1) = 1.9\ cm$$

One may conclude the following from the example.

a) For all practical purposes the terminal velocity is reached instantaneously by particles with diameters $d < 0.1$ mm.

b) The distance traveled in reaching 99% of u_t is minimal.

c) Particles with diameters $d > 0.1$ mm may be removed efficiently by gravitational settling in large updraft gas producers.

It is noted that an updraft rice hull gas producer with a diameter of 2 m at the top and with a gas output of 400 $m^3\ h^{-1}$ at 200°C will have a superficial gas velocity $u_s = 0.035\ m\ s^{-1}$. Because of the instantaneous terminal velocity of the particle (0.25 m s^{-1}), it is likely to remain in the gas producer. On the other hand, a small rice hull downdraft gas producer with a gas output of 12 m^3 h^{-1} at 600°C, will have a superficial gas velocity of 0.42 m s^{-1} if the gas is drawn off through a 10 cm diameter pipe or 0.15 m s^{-1} if the gas is drawn off through an annulus with an inner diameter of 25 cm and an outer diameter of 30 cm.

The above examples illustrate that the retention of small particles of d < 0.1 mm in the gasifier is also a function of the gas exit design. From the size distribution curve (Figure 6-21) we also can conclude that rice hull char contains about 48% - 60% of particles smaller than 0.1 mm. For larger particles with diameters d > 0.1 mm, Stokes theorem cannot be applied and the general set of equations relating the size of a sphere to its settling velocity must be written:

$$C_D \frac{d^2}{4} \rho_g \frac{u_t^2}{2} = \frac{\pi}{6} (\rho_s - \rho_g) g d^3 \qquad (10\text{-}5)$$

$$C_D = \frac{24}{Re} (1 + 0.15\ Re^{0.687}) \qquad (10\text{-}6)$$

$$Re = \frac{\rho_g u_t d}{\eta} \qquad (10\text{-}7)$$

The drag coefficient C_D is given by Equation (10-6) for the transitional region, $0.2 \leq Re \leq 500$, which is from Schillar and Nauman (1933). For the turbulent region $Re > 500$ C_D is roughly constant at $C_D = 0.44$ (Allen, 1975).

This system of equations must be solved by iteration. The viscosity, η, of the gas is strongly dependent on the temperature and is given in Figure 10-18 for a typical producer gas composition.

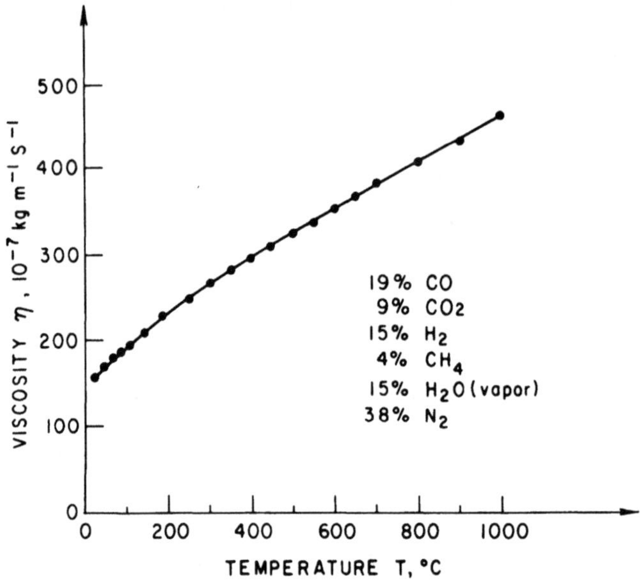

Figure 10-18. Producer gas viscosity as a function of temperature.

The terminal velocity of char particles smaller than 100 μm has been computed using equation (10-1) and is shown in Table 10-4. The assumptions made are that the raw gas exit temperature is 600°C and that the density, ρ_s = 2.0 g cm^{-3}.

Table 10-4. Terminal Velocity Versus Particle Size Diameter

Particle size, μm	100	75	50	25
Terminal velocity m s^{-1}	0.31	0.17	0.08	0.02

From Table 10-4 it can be concluded that removal of particles smaller than 100 μm is difficult to achieve by settling and requires a very low superficial gas velocity at the gas exit.

Air Injection

In the previous sections evidence has been represented to show that the method of drawing the air into the fuel bed is crucial for reliable rice hull gasification (Figures 6-11, 9-5, 9-7, 9-8). The majority of past downdraft gas producers had either wall tuyeres, a center tuyere, or a combination of the two. The following geometrical configurations have been tested with corncobs and rice hulls.

a) Wall tuyeres with and without throat (Figures 10-19 and 10-20).

b) Center tuyere with and without throat (Figures 10-21 and 10-22).

In order to avoid misinterpretation of the results, a series of tests were also conducted with a high quality gasification fuel (broken corncobs) of similar bulk density (≈ 200 kg m^{-3}). All tests were conducted in a 15 cm diameter refractory-lined gas producer equipped with either the vibrating or rotating grate shown in Figures 10-12 and 10-17.

The purpose of these tests was not to determine the arrangement which optimizes tar cracking. Instead the purpose was to determine the influence of the method of air entry on the gas composition. The results of the experiments are given by Kaupp (1982) in unpublished monthly reports to the Briggs and Stratton Corporation. The following was concluded:

a) None of the designs shown in Figures 10-19 through 10-22 are suitable for rice hull gas producers because the fuel bed is disturbed by the high superficial air velocity at the end of the nozzle.

b) The above designs work well for corncobs, a fact that has been known for decades.

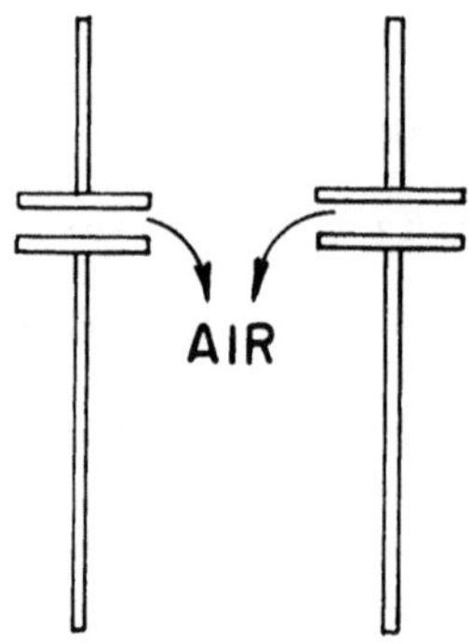

Figure 10-19. Wall tuyeres without throat, downdraft gas producer.

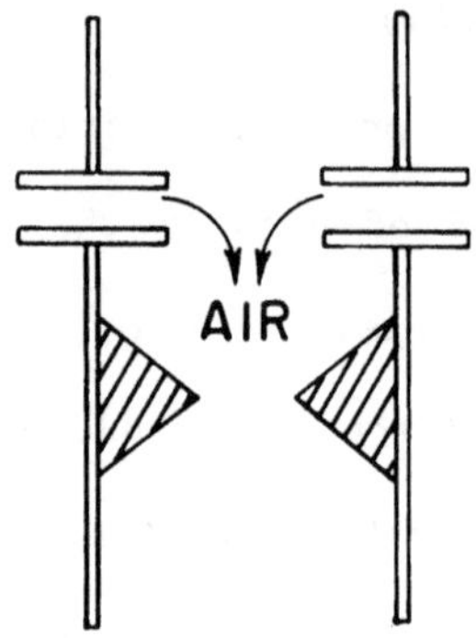

Figure 10-20. Wall tuyeres with throat, downdraft gas producer.

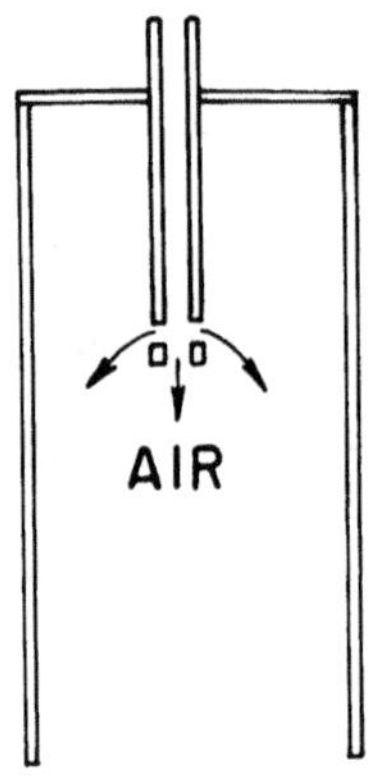

Figure 10-21. Center tuyere without throat, downdraft gas producer.

Figure 10-22. Center tuyere with throat, downdraft gas producer.

c) A small rice hull downdraft gas producer cannot have a throat, since the insignificant size reduction of rice hulls leads to blockage at the throat.

In order to avoid disturbance of the rice hull fuel bed, two different methods for drawing the air into the gas producer have been tested.

a) A continuous slot around the circumference of the gas producer which replaced the eight tuyeres (Figure 10-23).

b) Diffusion of the air through the fuel bed from above in an open top gas producer (Figure 10-24).

Based on experimental results with corncobs and rice hulls, it was concluded that diffusion of the air through the top is by far the most convenient way of introducing air without disturbing the fuel bed.

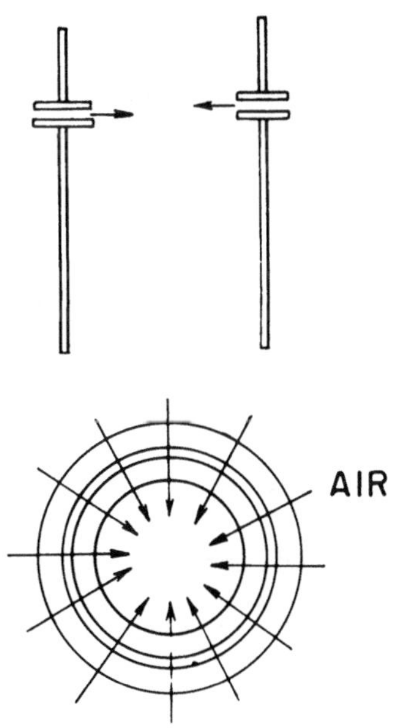

Figure 10-23. Continuous slot as air inlet.

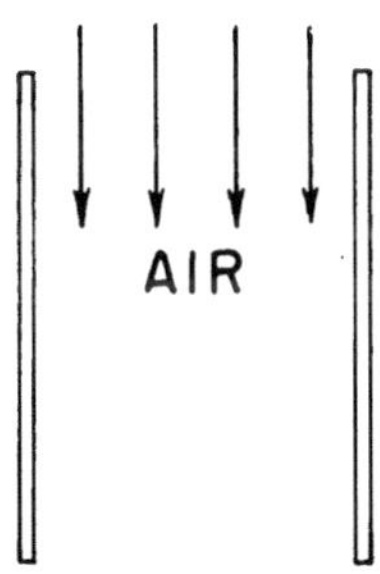

Figure 10-24. Open core air diffusion into the fuel bed.

Design of a Small (2 - 20 hp Mechanical Power Output) Rice Hull Gas Producer

Introduction

It was outlined in Chapter 3 that automotive gas producers which were fueled primarily with good quality charcoal, wood and coal. In addition large rice hull gas producers (60-200 hp) have been in operation for several decades prior to 1950 (Beagle, 1983). However, designing a small simple rice hull gas producer to power an internal combustion engine is a very challenging task for the following reasons:

a) The fossil fuel savings in small engines are relatively small. It is therefore necessary to have a design that can be inexpensively manufactured and that can be replaced after a few years of operation.

b) Contaminants in the raw gas (tar, solid particles) must be removed to a large degree within the gas producer itself because sophisticated gas cleaning equipment is either unavailable on a small scale or prohibitively expensive at this scale.

c) Any continuously moving part on a small rice hull gas producer must be driven by either an electric gear motor or a complicated drive arrangement involving the internal combustion engine. It is therefore highly desirable to avoid the incorporation of moving parts into the design.

Using these criteria, the approach of negative selection was used to show that the presently available designs for small, fixed-bed gas producers (crossdraft, downdraft, updraft) are not suitable for rice hull gasification. It is emphasized that it is possible to generate a combustible gas from rice hulls in some of these units at least in the laboratory. The objective of this work, however, has been to explore the behavior of rice hulls during thermal decomposition in order to establish parameters which would be helpful in improving designs.

In the previous chapters it has been outlined that the caking of rice hulls, the method of char removal and the method of injecting the necessary combustion air are crucial with regard to reliable operation of a gas producer. Based on these experimental results the following systems have been excluded and are considered unworkable for the gasification of rice hulls on a small scale.

a) Downdraft gas producers with a throat.

b) Downdraft gas producers with wall tuyeres or center nozzles or a combination of both.

c) Crossdraft gas producers.

d) Gas producers without a continuous ash removal system unless the units are batch fed.

e) Gas producers without a force feeding system unless the units are batch fed.

For instance, many versions of a downdraft unit were tested. These early models included the following:

a) Force feeding of rice hulls using an auger system with throat and center nozzle air inlet (see Figure 10-25).

b) Force feeding of rice hulls using an auger system without throat and center nozzle air inlet (not shown).

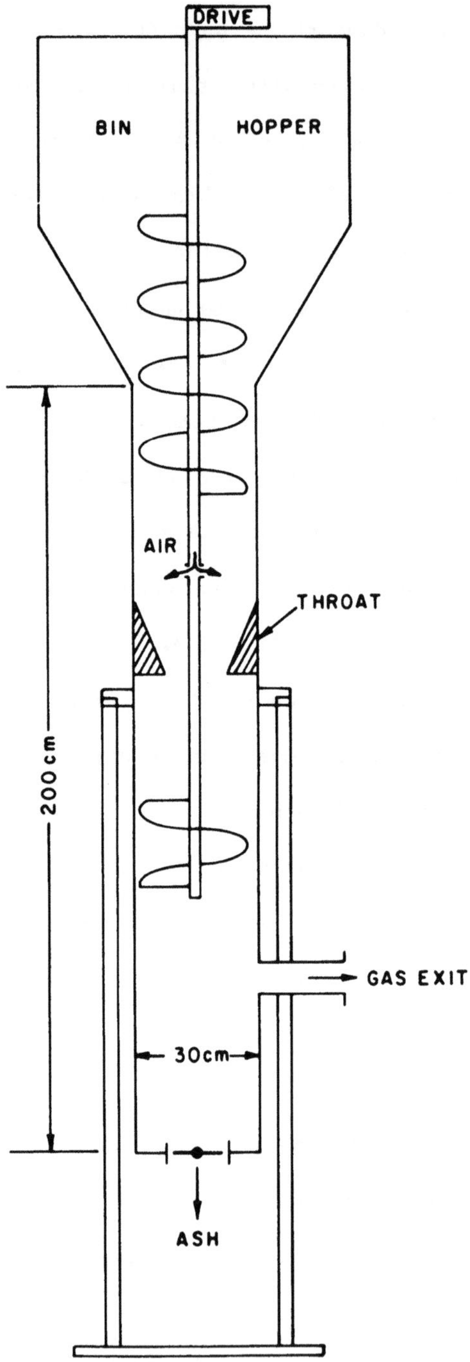

Figure 10-25. Downdraft gas producer with force-feeding system.

In both case the ash was removed by an auger as shown.

Although both systems could be operated for short periods of time (30 - 180 minutes), they have many design deficiencies as outlined below:

a) The center air nozzle generated caves in the fuel bed and induced slagging.

b) A force feeding system cannot move the rice hull char past the throat, which jams the auger (insufficient size reduction of rice hulls).

c) The system requires at least one external drive if the center pipe is water-cooled and two drives if the feeding and ash removal systems are separately driven.

It is noted that the raw gas generated had a sufficient higher heating value (3.7 - 4.4 MJ m^{-3}) and was fairly clean (the flame was blue to slightly orange).

Four other types of downdraft gas producers have been designed and tested. The results are presented in Chapters 6 through 9. All four prototypes have several common features:

a) The air is allowed to diffuse through the top.

b) All are equipped with a continuous ash removal system as described in Section 10.3.

c) The gas is drawn off above the grate.

d) High fire zone temperatures of > 1000°C should be maintained to ensure proper tar cracking and steam conversion.

The design shown in Figure 10-27 is the simplest version. It is equipped with a vibrating grate and is mounted on a free-swinging frame with the engine (see Figures 10-15 and 10-16). The design was tested extensively with corncobs and performed very well. The fire zone could be maintained anywhere desired within the fuel column (Kaupp, 1982). The rate of ash removal was controlled by adjustment of the distance between disk and firebox. Attempts to achieve continuous operation using rice hulls failed because of the caking of rice hulls (see Section 6.5). A schematic of the structure of the rice hull fuel bed before and after gasification is shown in Figure 10-26. The vibration induced by the engine did not break the carbon bonds established during the pyrolysis process because of the air gap between the fuel bed and the wall of the gasifier (see Figure 6-15). A slightly modified version, which was equipped with an effective and controllable wiper grate, did not perform any better because the wiper cannot remove rice hull char that does not move down onto the grate (Figure 10-28). A more complicated version is shown in Figure 10-29. With an auger mounted on top of the unit, the caked rice hulls could be pushed through the firebox column at a steady pace.

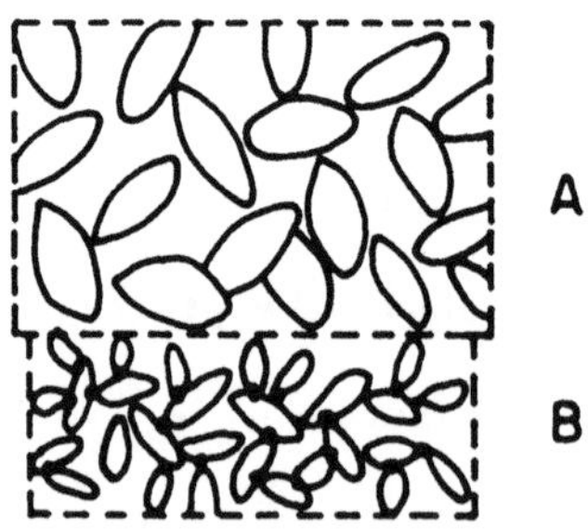

Figure 10-26. Rice hull fuel bed before (A) and after caking (B).

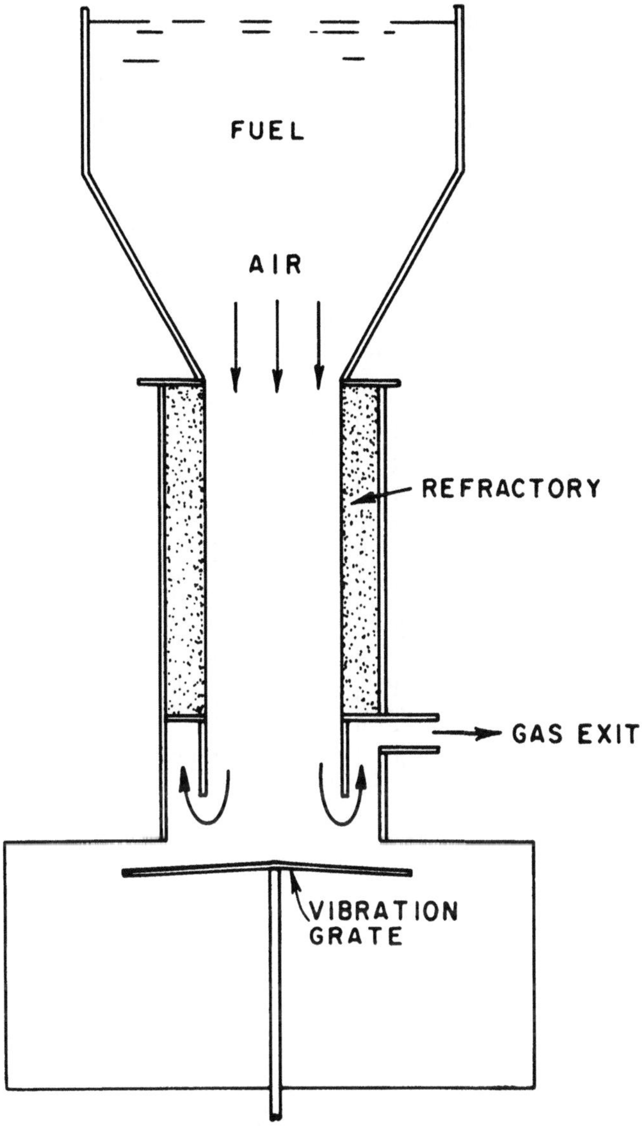

Figure 10-27. Gas producer with vibration grate and gravity flow.

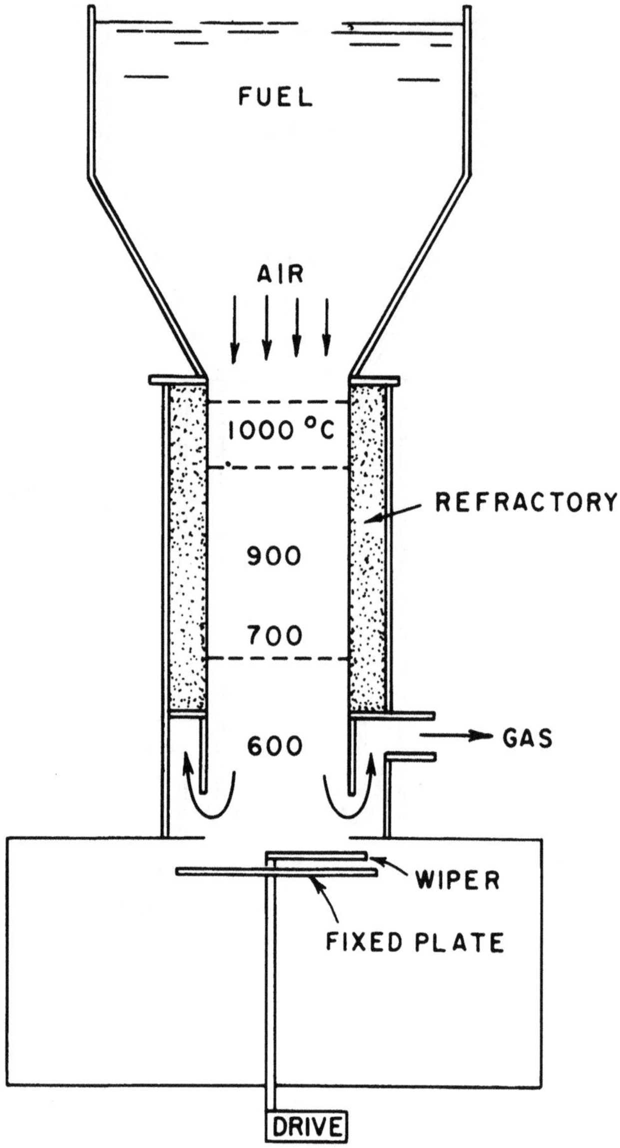

Figure 10-28. Gas producer with wiper grate and gravity flow.

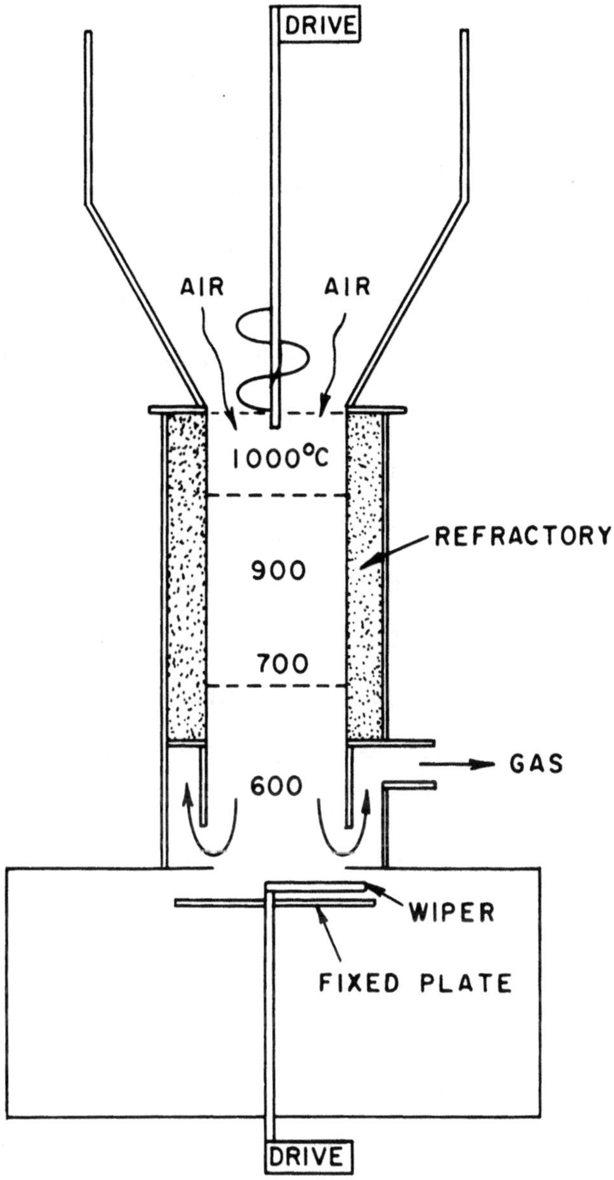

Figure 10-29. Gas producer with force-feeding system and wiper grate.

They are then removed from the grate by the wiper and discharged into the ash bin. Although this system does overcome the problems associated with caking, it introduces other complications. A careful coordination of the ash removal rate and the feeding rate becomes necessary. In addition, the feeding and ash removal rates must be coordinated with the gas output required by the engine. In order to avoid these difficulties and save one external drive, the grate was replaced with a hydraulic water seal (Figure 10-30). The water surface acts as a grate. It is noted that a rice hull column sitting on a water surface does not sink nor does it absorb a significant amount of water (≈ 1 inch/24 hours). Although the rice hull char is more porous than rice hulls and is therefore more likely to take up water and sink, the time required for this process is long when compared to the char velocity in the column. The water seal has the additional advantage of allowing continuous discharge of the ash into the water. The ash can then be removed from the water by hand. The water seal can also be used as an internal wet scrubber and cooler condenser. Despite the advantages of this system, it cannot be considered a practical solution to the problem of rice hull gasification on a small scale.

After testing all four prototypes over a period of 18 months, the concept of a "continuous" rice hull gas producer with a moving fuel column and a locally fixed fire zone was abandoned. Instead, a batch-fed unit with a moving fire zone and no ash removal during operation was designed and tested. The designs in Figures 10-27 through 10-30 show controlled char removal with the fire zone fixed in space while the air is diffusing in through the top as shown in Figure 10-31.

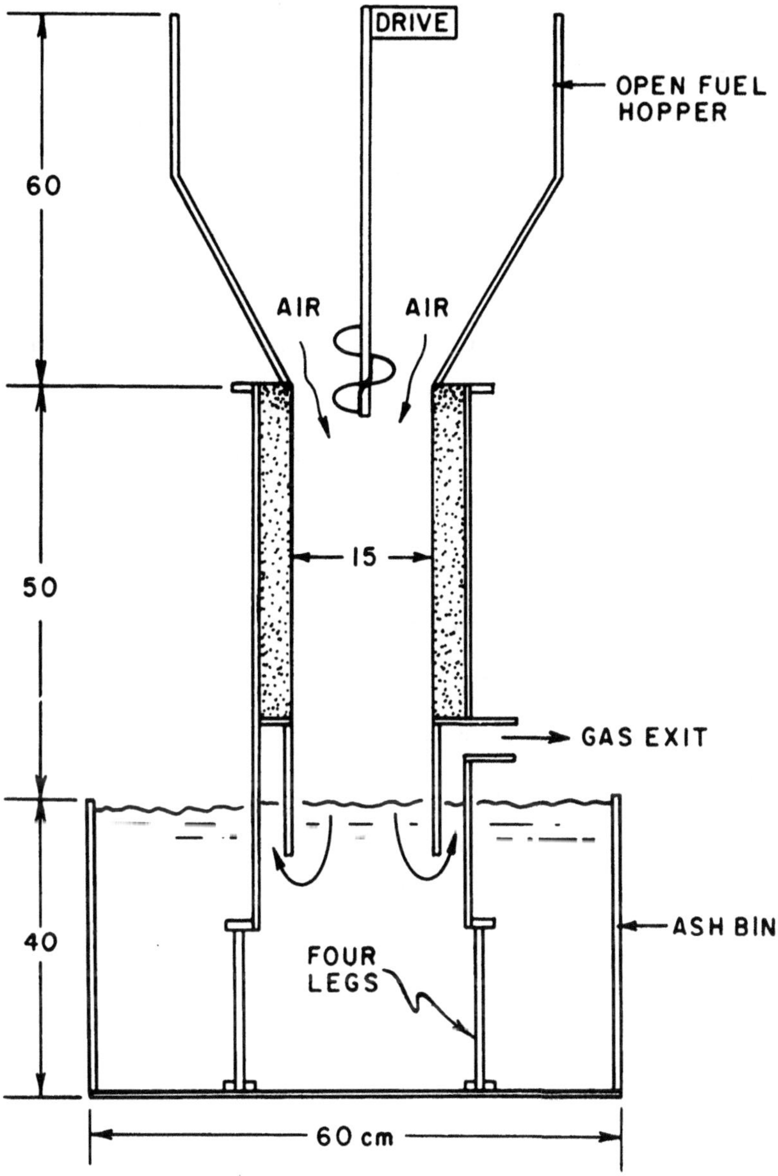

Figure 10-30. Gas producer with force-feeding system and water grate.

When gasifying rice hulls in a batch-fed unit, one cannot fix the position of the fire zone because the volumetric reduction of the fuel bed of 60% - 68% is insufficient. Two choices are available with regard to the operation of this type of a unit:

a) The fire is initiated at the top. The fire will then move down through the rice hull column leaving charred rice hulls behind. Below the fire zone will be uncharred rice hulls. This method results in a raw gas which is highly contaminated with tar because the gas does not pass through a hot column of char (Figure 10-32). (Similar to an updraft gas producer).

b) The fire is initiated at the bottom of the column. The fire zone will then move upwards against the air flow leaving a hot charred rice hull column behind.

This method (b) meets the necessary prerequisites for proper tar cracking and steam conversion. It also results in the generation of a carbon surface for CO_2 reduction to CO. Nevertheless, it is not obvious whether such a system can generate enough gas for continuous operation over a four-hour period. A detailed evaluation is given in the next section. It is noted that several researchers have simultaneously explored this new concept using wood fuel. Published literature concerning their work did not exist at the time of writing and information exchange was by personal communication (Bob Reines 1982, Tom Reed 1983, "Buck" Rogers Company 1983).

Open Core Gas Producer

The gas producer used in a set of six tests is shown in Figure 10-34. The gas exit of the test unit is located at the top instead of

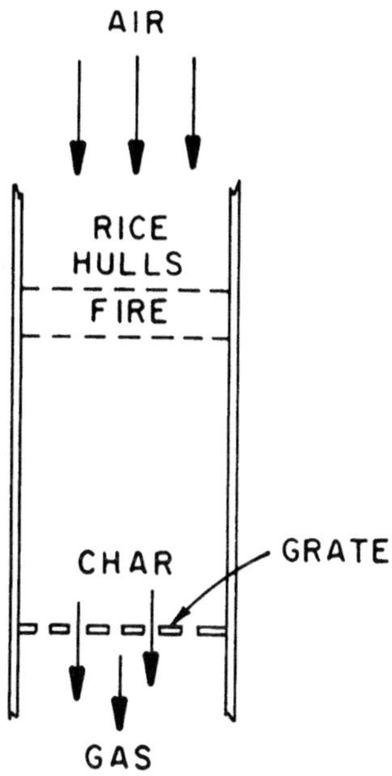

Figure 10-31. Fixed fire zone in an open core downdraft gas producer with continuous ash removal system.

in the middle. The middle position of the exit would be a better location because the dry gas cleaner would not be higher than the top of the producer itself. The design is based on the results reported in the previous chapters. The design features include the following:

a) A large annulus which serves as a heat exchanger and as a separation zone for coarse particles.

b) A steel restriction at the bottom which eliminates to some degree the wall flow shown in Figure 6-15.

c) A hydraulic seal, which allows one to empty the gas producer easily by lifting it out of the seal. The hydraulic seal can also be used as an internal wet scrubber and cooler for the hot producer gas.

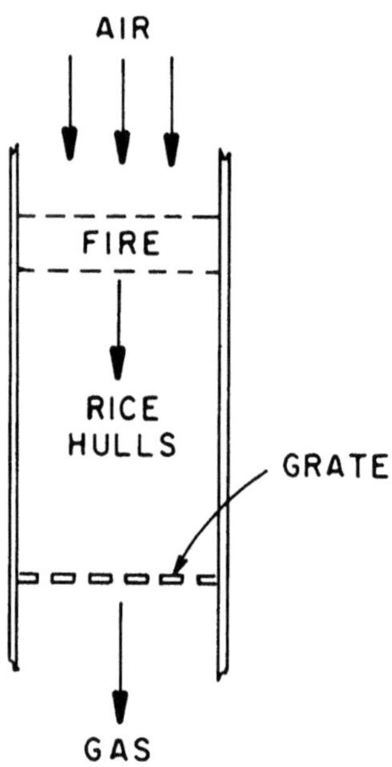

Figure 10-32. Batch-fed gas producer with top lighting.

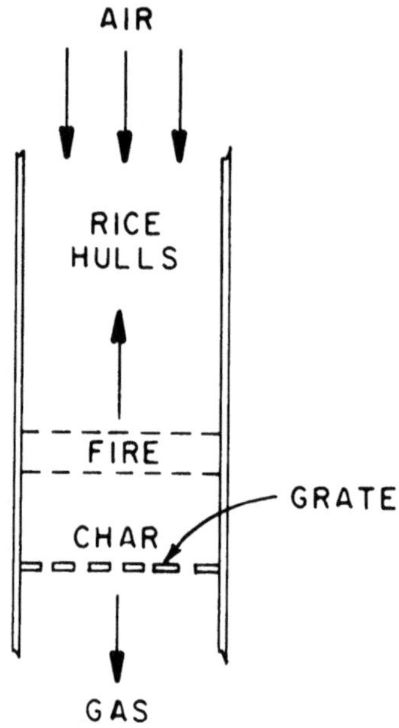

Figure 10-33. Batch-fed gas producer with bottom lighting.

d) A wet, sieve-plate scrubber followed by a dry bed scrubber of rice hull char.

Mode of Operation

The unit was filled with 20 - 40 grams of charcoal (wood or rice hulls) and was lighted by dropping a burning piece of paper through the top onto the charcoal. The necessary suction was provided by a suction pump, as a substitute for actual engine suction. When the entire cross section was glowing red, the column was filled with rice hulls. The purpose of the tests was to obtain all relevant operational data as a function of the gas flow, using a batch-fed operational mode (Figure 10-33).

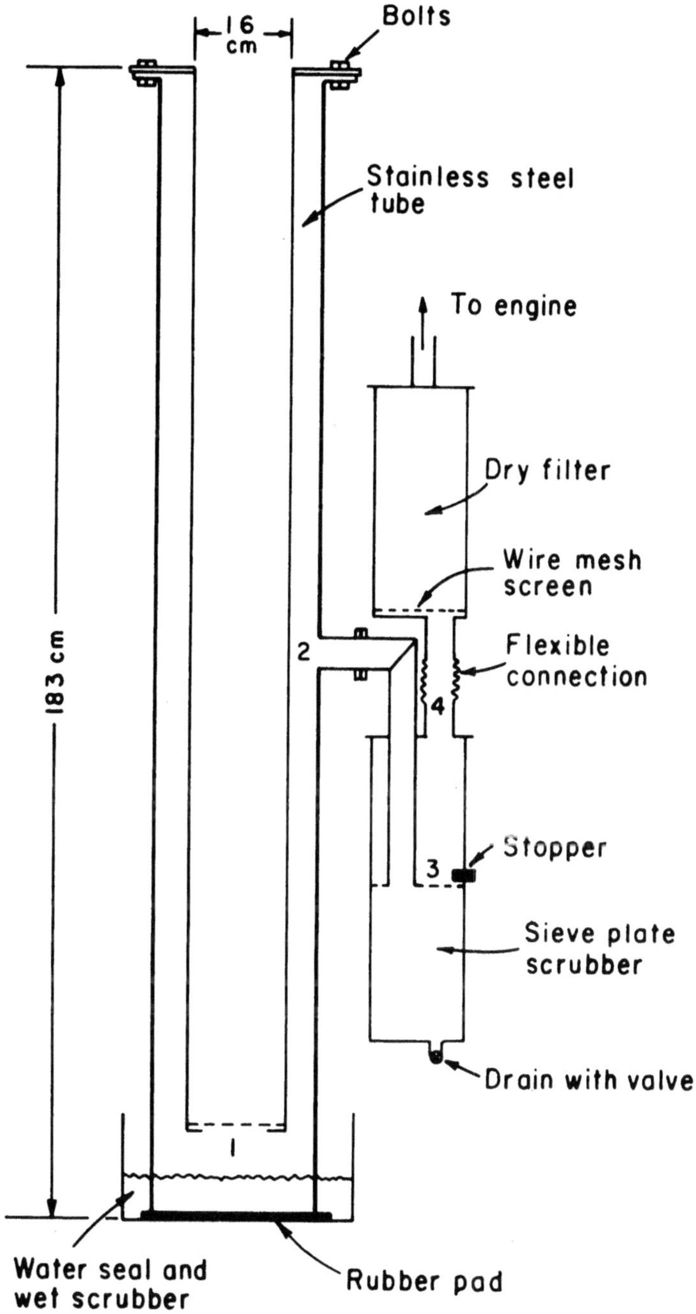

Figure 10-34. Batch fed gas producer without ash removal
1, 2, 3, 4 location of thermocouples.

From a theoretical and practical point of view, the following data are important with regard to rice hull gasification:

a) Air to fuel ratio.

b) Equivalence ratio.

c) Velocity of the upward moving fire zone.

d) Velocity of the downward moving rice hulls.

e) Gas composition CO, CO_2, THC, H_2, H_2O, N_2, O_2.

f) Higher heating value of the dry gas.

g) Superficial gas velocity within the fuel bed.

h) Time of operation for a 165 cm column.

i) Specific gasification rate.

j) Rate of rice hull consumption.

k) Degree of rice hull conversion.

l) Volume reduction of rice hulls.

m) Carbon conversion.

n) Efficiency of the process.

In addition, the efficiency of and the pressure drop across the gas cleaning train has been tested. It is noted that all results are given as a function of the gas flow rate which was varied between 2.76 and 9.9 $m^3\ h^{-1}$ (25°C, 1 atm). The gas flow rate was measured with a sharp-edged orifice plate. A very clean gas is necessary in order to measure the flow rate with this method. The fire zone temperature has a direct relationship with the gas flow rate. This is shown in Figure 10-51.

Air to Fuel Ratio

The air to fuel ratio is the amount of air used (by weight in kg) to gasify 1 kg of dry rice hulls. It can be computed once the total gas flow and the fraction of nitrogen in the dry producer gas are known. It was assumed that air consists of 21% oxygen and 79% nitrogen. The results are given in Figure 10-35. It is noted that the air to fuel ratio will vary according to the amount of producer gas demanded by the engine. It is not possible to determine an optimal air to fuel ratio without specifying the parameter to be optimized in a gas producer-engine system. The observed range of air to fuel ratios was between 1.28 and 2.1.

Equivalence Ratio ϕ

The meaning of the equivalence ratio ϕ as well as its importance has been discussed in Chapter 1 and shown in Figure 1-1.

From the test results (shown in Figure 10-35) a linear relationship between the air-to-fuel ratio and the equivalence ratio is apparent. It is noted that the equivalence ratio is computed by first determining the stoichiometric amount of air, i.e. the amount needed to combust the rice hulls completely, to CO_2 and H_2O with only the minerals as residue. The ultimate chemical analysis and the chemical formula of the rice hulls used for all tests is given in Table 10-5. One kg of these rice hulls requires 4.70 kg of air for stoichiometric combustion.

Table 10-5. Ultimate Chemical Analysis of Rice Hulls for Testing

Species	% weight, dry basis
C	39.00
H	4.90
O	33.97 by difference
N	0.53
Char residue	21.60

The measured moisture content was between 10.1% and 12% on a wet basis.

The equivalence ratio at which the higher heating value of the gas was maximized was 0.32. This is shown in Figure 10-41.

Velocity of the Fire Zone and Velocity of the Rice Hull Fuel Bed

Because no ash removal takes place, the high ash content of the rice hulls results in an upward moving fire zone. On the other hand, the volume reduction of rice hulls following combustion and their caking behavior result in a downward movement of the rice hull fuel bed (Figure 10-26). The test was terminated when the entire 165 cm long column was filled with char and the fire zone had reached the very top of the unit. Both fuel and fire zone velocities are important when considering long-term operation and when considering the use of a forced feeding system as shown in Figure 10-30.

The fire zone velocity is defined as the ratio of the length of the fuel column to the total operating time. The downward fuel velocity was measured two to three times during each run. The results are given in Figure 10-36.

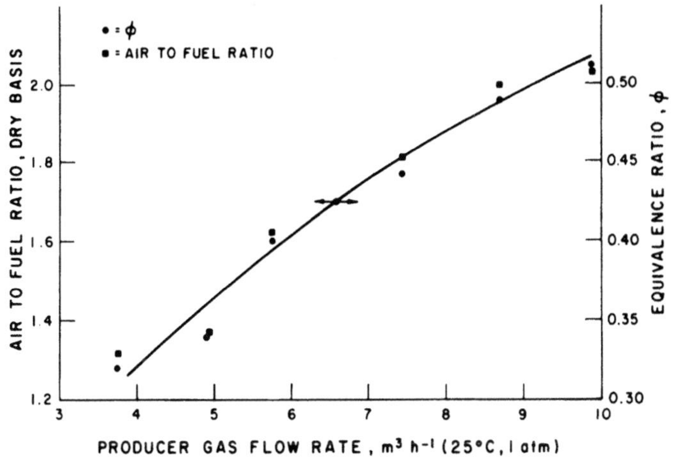

Figure 10-35. Air to fuel ratio of the gas producer as a function of the gas flow rate.

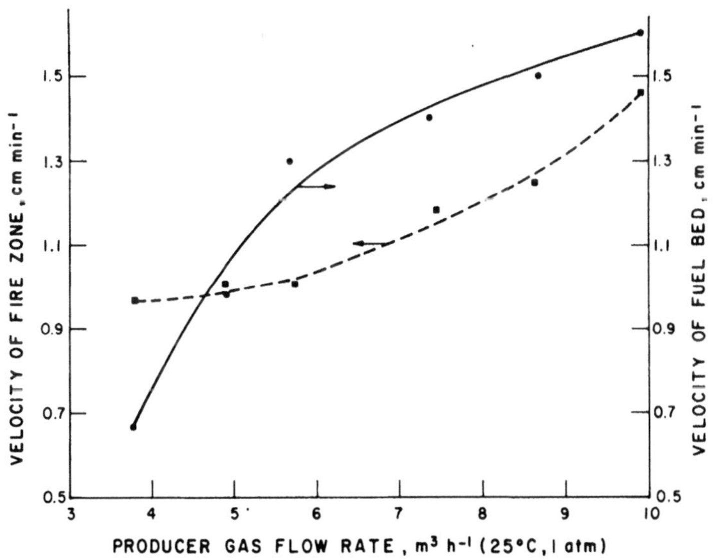

Figure 10-36. Fire zone velocity (up) and fuel bed velocity (down) as a function of the gas flow rate.

Gas Composition

From the theoretical considerations discussed in Chapter 5, the expected behavior of the various components of the gas is as follows:

a) As a function of the gas output, the volume fraction of CO and H_2 will reach a maximum within the measured range. The fraction of CO and H_2 will decrease with decreasing flow rate (lower ϕ) because the rate of steam conversion and CO generation by Equations (5-2 and 5-4) are a strong function of temperature (Figure 9-12).

   A similar decrease in CO and H_2 will occur at higher flow rates (higher ϕ) due to internal oxidation and to the water-shift reaction. This typical behavior of CO and H_2 is shown in Figure 10-37. The combined CO and H_2 content was maximized at a flow rate of 7.5 $m^3\ h^{-1}$ or a ϕ of 0.44. However, this does not mean that the chemical energy content of the gas is maximized under these conditions (see Figure 10-41).

b) The methane content of the gas decreases with increasing gas flow rate. As outlined in Chapter 5, the methane content will decrease drastically with increasing temperatures. This is shown in Figure 10-38. Due to the unique mode of operation, which results in a long rice hull char column downstream from the fire zone, lower flow rates and an increased methane content did not result in a measurably higher tar content in the raw gas. It is noted that the amount of total gaseous hydrocarbons was measured (THC). However, it is assumed that only traces of C_2H_m and C_3H_m are present (see Figure 5-22).

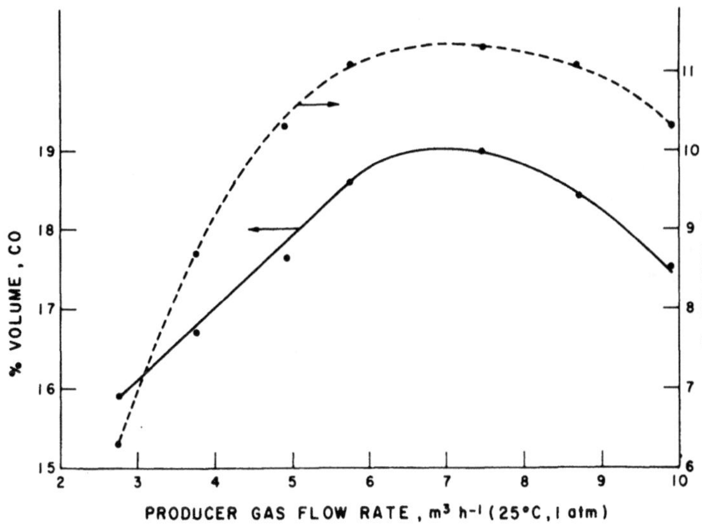

Figure 10-37. CO and H_2 content of the dry gas as a function of gas production rate.

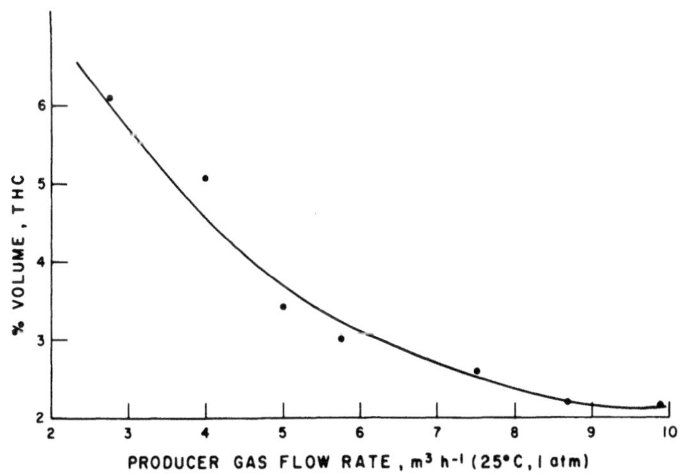

Figure 10-38. THC content of the dry gas as a function of gas production rate.

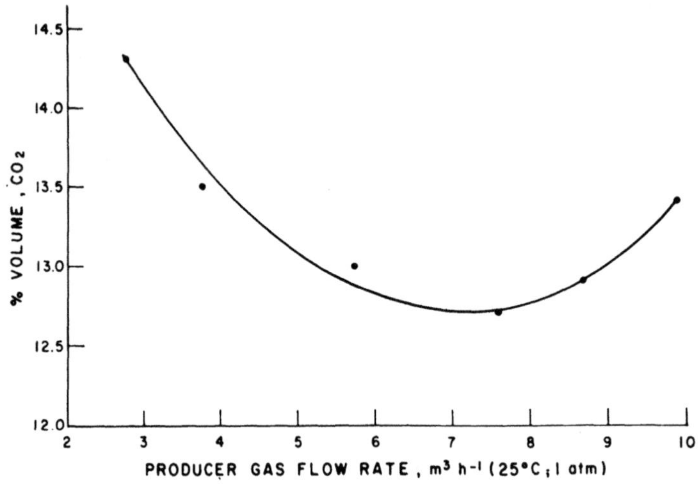

Figure 10-39. CO_2 content of the producer gas as a function of the gas flow rate.

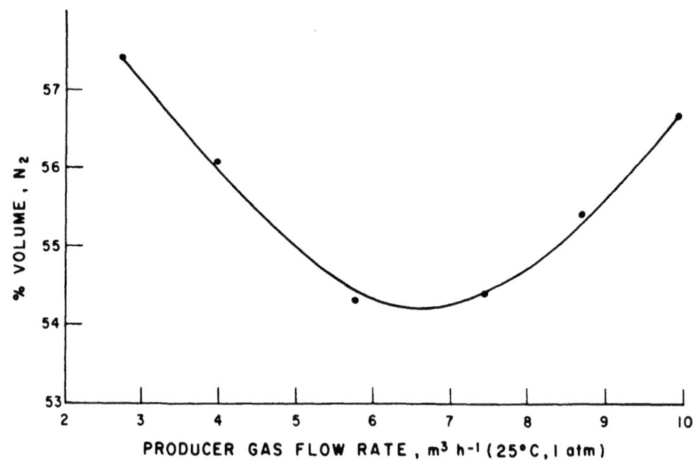

Figure 10-40. N_2 content of the producer gas as a function of the gas flow rate.

c) The amount of CO_2 in the gas will be minimized within the range of operation. An increase in CO_2 must be expected at low and high gas flow rates due to insufficient reduction of CO_2 and internal oxidation respectively. This is shown in Figure 10-39.

d) The volume fraction of nitrogen in the gas is measured by difference and therefore must follow the CO_2 pattern. This is shown in Figure 10-40.

All results are based on gas samples that were oxygen free. It is emphasized that during the start-up period (5 - 20 minutes) oxygen is present in the gas stream. This period can be considerably shortened by allowing the fire zone to develop over the entire cross section before adding the rice hulls. A rice hull fire zone tends to develop very slowly in a horizontal direction if there are unburned rice hulls in the vertical direction. Once the fire zone was fully developed, no oxygen could be detected in the gas stream.

A summary of the range of gas compositions as a function of the gas flow rate is given in Table 10-6.

Table 10-6. Dry Gas Composition as a Function of the Gas Flow Rate

Flow rate	CO	CO_2	THC	H_2	N_2	M_g
$m^3\ h^{-1}$			% volume			
2.75	15.9	14.3	6.1	6.3	57.4	27.9
4.97	17.6	13.1	5.1	9.6	54.6	27.0
5.74	18.6	13.0	3.0	11.1	54.3	26.8
7.46	19.0	12.7	2.6	11.3	54.4	26.8
8.68	18.4	12.9	2.2	11.1	55.4	26.9
9.9	17.5	13.4	2.2	10.3	56.6	27.2

The same mode of operation has been extensively tested using corncobs in the unit shown in Figure 10-27. For instance, a continuous 7-hour test with corncobs resulted in 22% CO and 15% H_2 (Kaupp, 1982). It is to be noted that the relatively low CO and H_2 content of the gas is not due to the mode of operation or to the design but is due to the structure of rice hulls and the make up of a rice hull fuel bed.

Higher Heating Value of Producer Gas

The higher heating value of the gas was computed based on the following heating values of the individual gas species at 25°C:

CO	11.57 MJ m^{-3}
H_2	11.69 MJ m^{-3}
CH_4	36.41 MJ m^{-3}

The results are given in Figure 10-41. It is emphasized that the behavior of the higher heating value as a function of the gas output is very typical of that for downdraft producer gas operation in general. The following characteristics of this behavior are noted:

a) Maximization of the CO or H_2 contents does not result in maximization of the chemical energy content of the gas. For instance, the CO content can be as high as 32% (by volume) under conditions favoring the water shift reaction in such a unit (Kaupp, 1982). However the heating value will drop sharply because almost no CH_4 and very little H_2 (< 5%) will exist under such conditions.

b) It is noted that even small increases in CH_4 at lower flow rates can offset moderate losses in H_2 and CO because CH_4 has three times the heating value of CO or H_2.

c) Operation at very low gas flow rates also means low fire zone temperatures. Thus, the chemical energy content of the gas will decrease again in this range because the large decreases in CO and H_2 cannot be offset any longer by increases in CH_4. The chemical energy content of the gas was maximized at a flow rate of 4 $m^3\ h^{-1}$ or a ϕ of 0.32 whereas the combined H_2 and CO content was maximized at a flow rate of 7.5 $m^3\ h^{-1}$ or an equivalence ratio of 0.44.

The higher heating value of the gas lies between the 3.0 MJ m^{-3} reported for the Chinese downdraft gas producer (Section 10.2) and the 6.72 MJ m^{-3} reported for the Balestra unit (Section 10.1).

Superficial Gas Velocity, v_s

The superficial gas velocity, v_s, within the rice hull column is defined as the gas velocity that would result if the cylinder was empty.

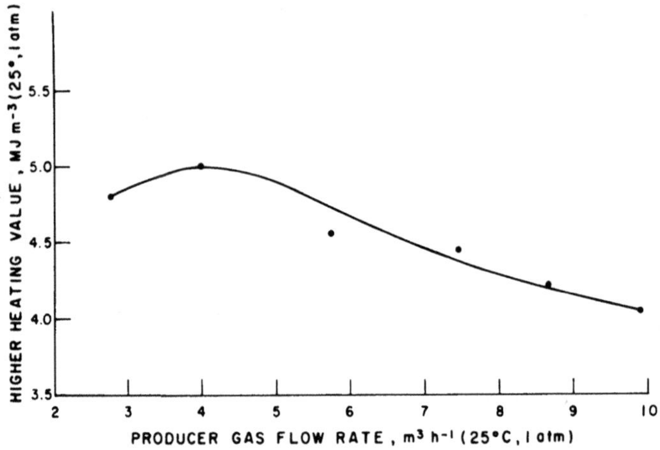

Figure 10-41. Higher heating values of the producer gas.

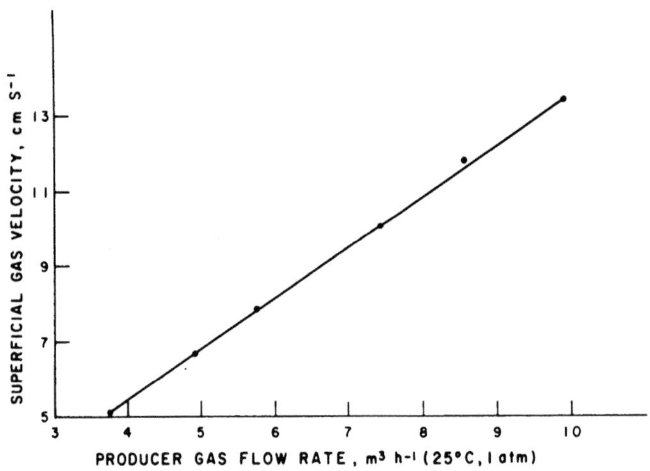

Figure 10-42. Superficial gas velocities (25°C) as a function of gas flow rates.

Its values are given in Figure 10-42 for the range of producer gas outputs. It is noted that the numbers are computed using a temperature of 25°C. From the graph it can be seen that high gas outputs of more than 6 $m^3\ h^{-1}$ lead to superficial gas velocities that can slightly disturb the fuel bed, as shown in Figure 6-18.

The actual gas velocity, v_a, at any point in the gas producer can only be computed if the temperature and void space fraction, ε_β, is known at this point. The relation between v_s and v_a is then given as:

$$v_a = v_s \frac{T_1}{(298)(\varepsilon_\beta)} \qquad (10\text{-}8)$$

where T_1 is the fuel bed temperature.

This equation assumes that $\lim_{V \to 0} \varepsilon_\beta = \varepsilon_a$ where ε_a is the area void space and V is a control volume of cross-sectional area A and height Δz with $\Delta z \to 0$.

Operating Time for a 165 cm Column

Due to the high ash content of rice hulls a short to moderate operating time can be expected at various flow rates, as shown in Figure 10-43. During the first 5 - 15 minutes of operation, the gas is not suitable for engine use. However, the gas composition then remains stable up to the point where the fire zone reaches the very top of the unit. This system can be made continuous at additional expenses by adding a force feeding system and a continuous ash removal system.

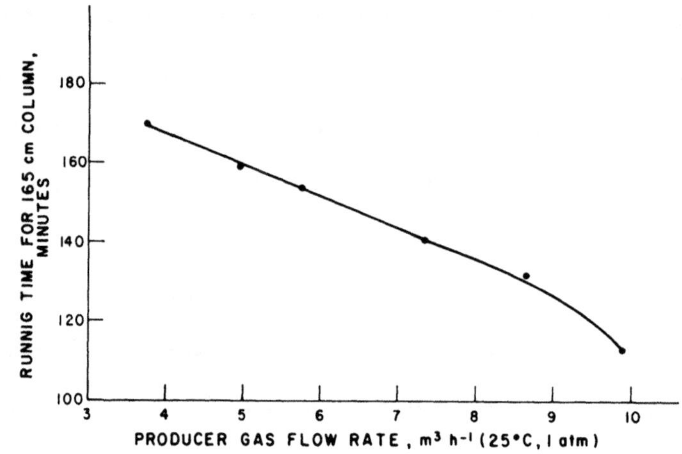

Figure 10-43. Operation time for a 165 cm column.

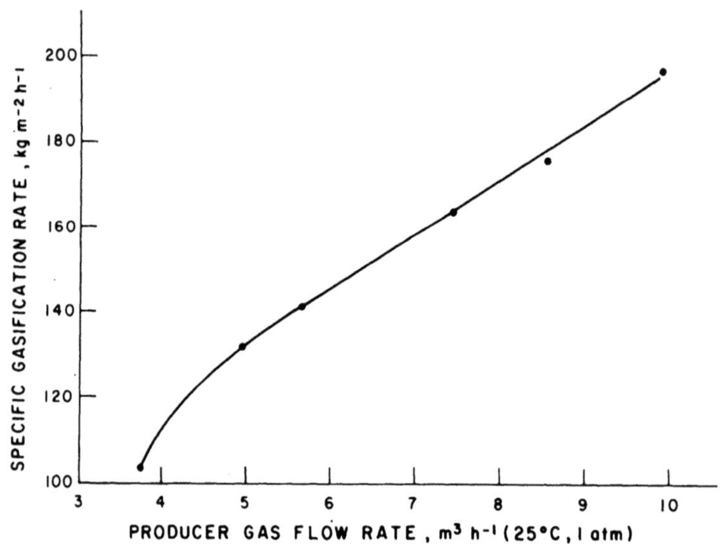

Figure 10-44. Specific gasification rate.

Specific Gasification Rates

The specific gasification rate is an indicator of the size of the gas producer needed for a given output. This parameter is strongly dependent on the type of gas producer and on the fuel. For instance, values of 100 - 300 kg m^2 h^{-1} are considered normal for large coal updraft gas producers (Gumz, 1952), while values of 225 kg m^{-2} h^{-1} to 5000 kg m^{-2} h^{-1} are not uncommon in downdraft gas producers operated with wood, corncobs and charcoal (Jenkins, 1980; Groeneveld, 1982). It is pointed out that the assessment of the specific gasification rate in a downdraft gas producer with a throat depends on which cross-sectional area is used as a reference. Because this is not always specified, there is a high uncertainty of 200 - 300% in reported values. From Figure 10-44 it can be concluded that the tested open core downdraft gas producer has a low specific gasification rate for rice hulls of 110 - 210 kg m^{-2} h^{-1} similar to that found in coal gasifiers.

Rate of Rice Hull Consumption

The rate of rice hull consumption, on a dry basis, is shown in Figure 10-45. A range of 2.11 to 4.05 kg h^{-1} for the tested gas flow rates was observed. It is noted that the volume of producer gas generated per kg of rice hulls gasified increased from 1.78 m^3 h^{-1} to 2.44 m^3 h^{-1} with increasing flow rate but with a reduction in energy content.

It is assumed that the producer gas can be converted into mechanical shaft power with an overall efficiency of 15% in a small gasoline engine. The resulting range of horsepower outputs (1.0 to 2.2) is also shown in Figure 10-45. Because of the lower volumetric efficiency when

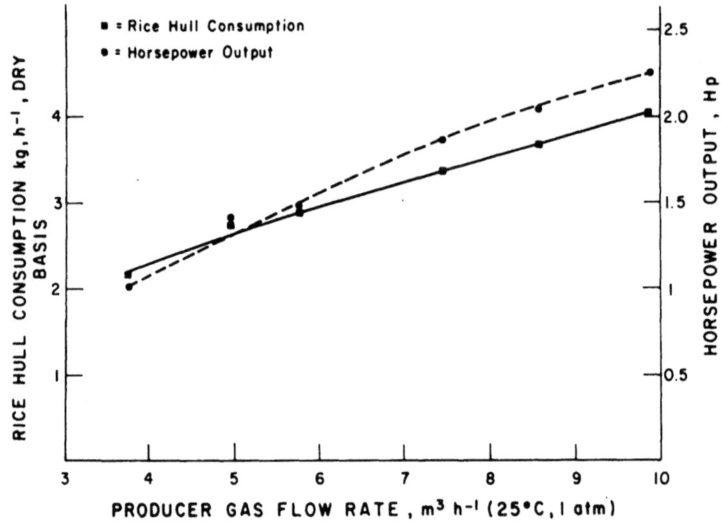

Figure 10-45. Rate of rice hull consumption, and computed horsepower output at 15% engine efficiency.

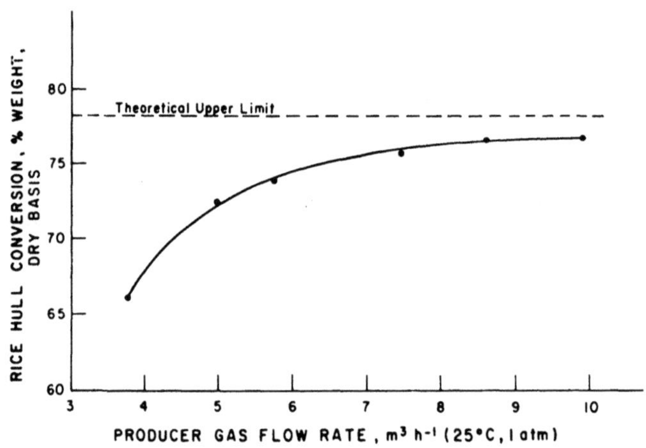

Figure 10-46. Degree of rice hull conversion.

operating on producer gas, and the lower energy density of the gas, a gasoline engine must have a gasoline rating at least twice the desired power output when operated on producer gas. Based on evaluations of past systems, 1.75 - 3.00 kg of rice hulls (as received) were necessary to generate one horsepower-hour. The results given in Figure 10-45 show a dry rice hull consumption of 1.76 - 2.0 kg per horsepower hour and are therefore in agreement with past data.

Degree of Rice Hull Conversion

Due to the high ash content of rice hulls, one has to cope with large amounts of residue following gasification. The extent to which rice hulls are reduced in weight following gasification is shown in Figure 10-46. The theoretical limit of a 78.4% conversion or a 21.6% residue retention by weight is based on the mineral content. Not all the minerals stay in the gas producer. The bulk of the mineral matter has a very high melting point and therefore an even higher boiling point. However, some fractions are indeed boiled off at lower temperatures. The characteristic white spots on the inside walls of a rice hull gas producer and piping system suggests that at least some silicon is given up as SiO and is later oxidized to SiO_2, which is in the form of a very fine powder. Based on the assumption that no minerals are leaving the gas producer, a weight conversion of 66.3% to 76.8% takes place. It is pointed out that this represents a conversion of 84.6% to 97.9% of the combustible matter C, H, O, N. The degree to which carbon was converted is shown in Table 10-7.

Volume Reduction of Rice Hulls

Because of the mode of operation, the volume of residue following gasification is identical to the volume of the inner steel cylinder. In fact, the residue is a loosely connected cake of rice hull char. Based on the experiments, a volume reduction of only 42% to 57% can be expected (Figure 10-47). Gasification of rice hulls, therefore, is certainly not a solution to rice hull disposal because 58% to 43% of the original volume (undisturbed) is present after gasification in the form of char. It has been previously pointed out that when the fuel bed is stirred or when the rice hulls are discharged through a rotating grate, 32% - 37% of the initial volume will be occupied by the residue.

Carbon Conversion

The term carbon conversion refers to the degree to which the carbon in the fuel has been converted into gaseous and liquid products. The amount of solid carbon carried out in the gas stream is negligible. The ultimate chemical analysis of the rice hulls was given in Table 10-5. The ultimate chemical analysis of the residue as a function of the flow rate is given in Table 10-7. If x_f is the weight fraction of carbon in the rice hulls and x_c is the weight fraction of carbon in the residue then the carbon conversion τ is given in percent on a dry basis by:

$\tau = 100 \left(1 - \frac{m_c x_c}{m_f x_f}\right)$ where m_f and m_c is the dry mass of the rice hulls and their residue respectively.

Table 10-7. Residue and Its Chemical Ultimate Analysis

Flow Rate $m^3\ h^{-1}$	Carbon Conversion %	Species % weight				
		C	O	H	N	Mineral
3.76	67.4	37.78	3.78	0.65	0.29	57.5
4.97	81.1	26.78	3.97	0.16	0.19	68.9
5.74	85.0	22.43	3.31	0.29	0.17	73.8
7.46	87.9	19.34	2.15	0.27	0.14	78.1
8.68	90.3	16.26	2.51	0.22	0.11	80.9
9.9	90.6	15.75	1.55	0.21	0.09	82.4

From Table 10-7 one can conclude that only 67.4% to 90.6% of all the carbon in the fuel is utilized. It is noted that even at the higher flow rates (and higher temperatures), moderate amounts of volatiles are still present in the residue. The residue obtained from the test with the lowest flow rate has a chemical composition similar to the residue obtained from low temperature pyrolysis (see Figure 8-7 and 8-8).

Efficiency of the Process

The efficiency of a gas producer is defined here as:

$$\eta = \frac{\text{chemical energy output, cool dry gas}}{\text{energy input from rice hulls}}$$

Because the total volumetric flow was measured as well as the total consumption of rice hulls, η can be computed. The following assumptions were used:

a) The higher heating value of rice hulls is 15.3 MJ kg^{-1} on a dry basis (measured).

b) The higher heating value of the gas is used for the chemical energy output.

The efficiencies, which vary between 55.8 and 66.5%, are shown in Figure 10-48. This efficiency is most commonly called the cold gas efficiency of a gas producer. It is emphasized that the efficiency curve must reach a maximum over a range of gas outputs. At low gas flow rates too much carbon remains in the rice hulls. At high flow rates, the carbon utilization is quite good (see Table 10-7), but the chemical energy content of the gas deteriorates (see Figures 5-23 and 10-41), with a higher sensible heat fraction.

Gas Cleaning Train

Introduction

The difficulties encountered in the cleaning of producer gas have hampered the widespread use of producer gas as an engine fuel. It is reported that the generation of producer gas from biomass is not a difficult task. However, much effort is required to generate producer gas in a form which does not require complicated and expensive gas cleaning equipment and operates with a low pressure drop.

The present lack of reliable data with respect to the performance of gas cleaning equipment is mainly due to measurement difficulties and to the need for expensive monitoring equipment.

As part of this research, several very different methods of gas cleaning have been tested and are reported by Kaupp (1982). A gas cleaning train in general must perform several tasks:

a) Cooling of the gas to near ambient temperature.

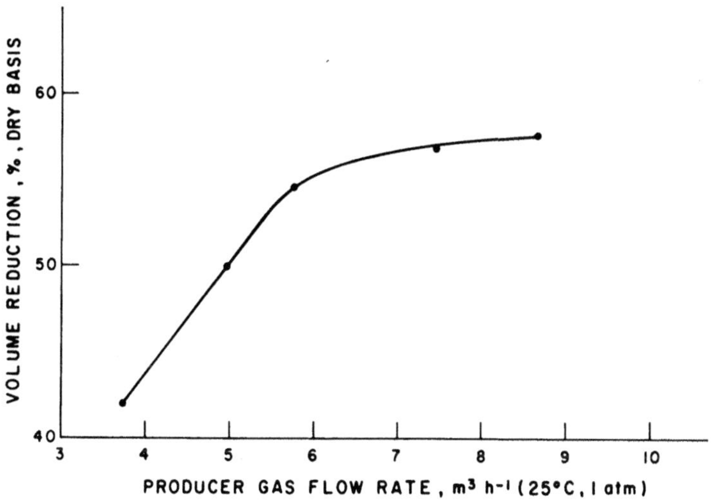

Figure 10-47. Rice hull volume reduction.

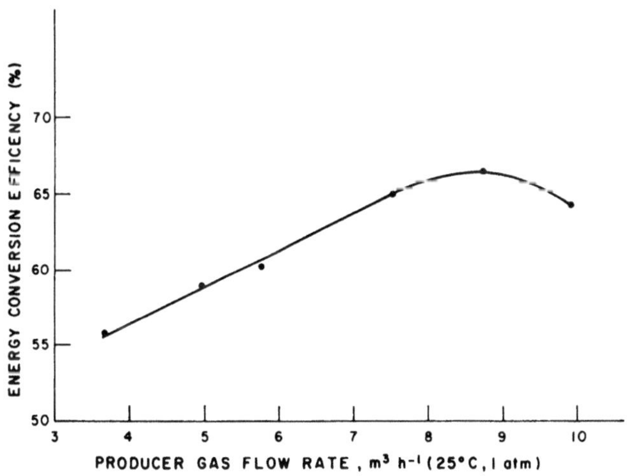

Figure 10-48. Efficiencies of the process.

b) Removal of solid particles from the gas stream.

c) Removal of tar vapors and steam from the gas stream.

It is therefore somewhat unreasonable to expect a single device to accomplish these different tasks.

During the heyday of automotive gas (charcoal) producers, a solid particle load of 5 - 15 mg m^{-3} of gas at ambient conditions was an accepted maximum standard. It was claimed that higher solid particle loads will result in decreased engine life. It is noted that the raw gas carries solid particles, tar vapors and steam. No regulations existed with regard to the allowable amounts of tar vapors and steam in the gas stream.

The following systems were tested (Kaupp, 1982):

a) Dry packed bed filter (rice hull char).

b) Dry cyclone.

c) Wet cyclone.

d) Sieve plate scrubber in combination with a dry packed bed filter of rice hull char.

Only option d), which has proven to be a highly efficient combination, will be analyzed in detail.

Sieve Plate Scrubber and Dry Packed Bed Filter

The system and its dimensions are shown in Figure 10-49. Cooling alone of the gas by passing it over or through a water film is not adequate for efficient gas cleaning. One of the foremost problems in gas cleaning is the removal of very fine particles and the separation of tar mist from the gas stream. The difficulty lies in the condensation of

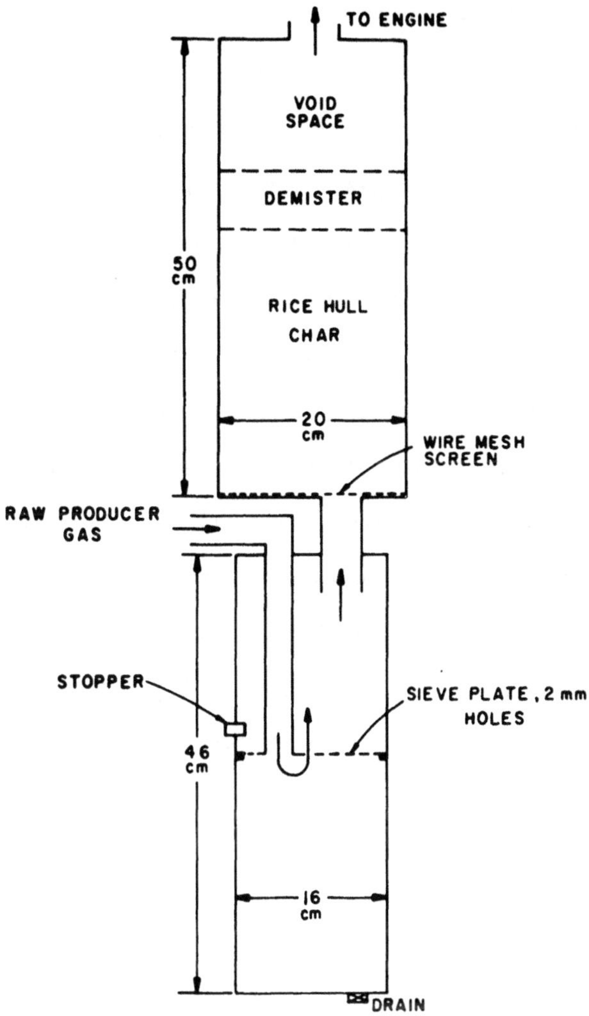

Figure 10-49. Gas cleaning filter train.

the higher hydrocarbons (tar) at temperatures between 100°C and 400°C. This tar mist will still remain in the gas stream and will partially adhere to the wall surfaces and to dust in the gas. These tar vapors and solid particles give the gas a foggy appearance.

There is considerable disagreement over how clean producer gas must be in order to give an internal combustion engine long and trouble-free operation. However, it is a simple fact that "clean" producer gas, when observed through a glass viewing port, is completely invisible to the human eye. This rule of thumb immediately separates the good gas producers and gas cleaning trains from the poorly functioning systems. It is noted that "clean" air still contains a sizeable amount of particles as shown in Table 10-8. Comparing these figures with the requirement that producer gas has a maximum of 5 - 15 mg m^{-3} (at 25°C), it is clear that reducing the particulate content of the gas to such low limits is a formidable task.

Table 10-8. Particulate Content of Air (SERI)

Air In	mg m^{-3}
Rural areas and suburbs	0.5 - 1
Cities	2
Industrial centers	4
Streets with heavy traffic	20
Dusty highways, farm work with tractors	over 200

The purpose of this work was to design a simple, inexpensive, and reliable gas cleaning apparatus that does not generate an excessive pressure drop. The excessive pressure drop caused by most gas cleaning trains is a serious problem. Most gas cleaning devices, such as fabric filters and packed bed filters, are able to clean the gas to a certain degree, but the excessive pressure drop caused by these filters makes it necessary to install a pump between the gas producer and the engine. In most installations, the engine itself is used to overcome the entire pressure drop across the gas producer and gas cleaning train, which can cause a severe drop in the volumetric efficiency of the engine and therefore a substantial power drop (Kaupp, 1982). Taking all this into consideration, a sieve plate scrubber followed by a dry bed filter of rice hull char was chosen.

It is emphasized that efficient tar and fine particle removal requires a highly atomized water spray and therefore a large surface area and turbulent motion to impinge the contaminants. In order to avoid the need for a high pressure water line and a pump, the gas itself was used to induce turbulent mixing of gas and water.

Designing sieve plate scrubbers for a given gas flow rate and pressure drop is an old technology and treated in detail by Treybal (1980) and Calvert (1972). The tested scrubber, however, is not a standard design and has no water flowing across the plate. The hole diameter, d_o, is 2 mm. The holes are located at the corners of equilateral triangles at a distance d = 4 mm between centers. The ratio of total are, A, to hole are, A_o, is then given by

$$\frac{A_o}{A} = 0.907 \left(\frac{d_o}{d}\right)^2$$

(Treybal, 1980). Therefore the area available for the gas flow is 22.6% of the total cross-sectional area. In order to avoid excessive wall flow and large bubbles, the sieve plate should be pressed against a steel ring to get a tight fit at the wall. The entire gas cleaning train was made from plexiglass in order to observe the motion of water and gas. A wet sieve plate scrubber can be operated without a continuous water flow for small units although continuous water flow is preferred because it enables cooling of the gas to near ambient temperature. The water temperature of the wet scrubber should never reach the boiling point because the repulsive forces of boiling water will actually repel the particles instead of collecting them.

The tests reported here were conducted without a continuous water flow. This is one reason why the space below the sieve plate is relatively large even though this part does not contribute to gas cleaning. The apparatus was also tested with a continuous water flow during the gasification of corncobs in the unit shown in Figure 10-28 (Kaupp, 1982).

Experimental Procedures and Results

The principle of operation was as follows:

Before starting the gas producer the sieve plate scrubber is filled with water to a level 1 - 3 cm above the sieve plate depending on the desired pressure drop and on the gas flow rate. After the gas producer was started, the incoming raw gas is finely divided by the sieve plate into many small bubbles. Above the sieve plate a highly turbulent mixture of water and gas (froth) is formed which provides an excellent environment for impinging tar droplets and solid particles. At the same time the gas is cooled. When the gas leaves the foam bath it contains entrained water mist which will be partially eliminated in the packed bed filter. The packed bed filter is purposely located above the wet scrubber in order to allow drainage of the collected water back into the wet scrubber. As the saturated gas cools, the moisture in the gas condenses and creates a film covering the rice hull char, which provides additional particle removal in the dry bed. Gas sampling was done before and after the gas cleaning train. The continuous gas sampling before the gas cleaning train was used to determine the moisture content of the raw gas at the gas exit and the gas composition. The continuous gas sampling after the gas cleaning train was used to determine the moisture content and the solid particle content of the gas. Consequently, the following measurements were taken:

a) Moisture content of the raw gas at the gas exit of the gas producer.

b) Moisture content of the gas after leaving the gas cleaning train.

c) Particle content and liquid higher hydrocarbon content condensing out above 105°C.

d) Pressure drop across the wet sieve plate scrubber.

The tar and solid sampling was done according to EPA Method #5, which is used to sample the contaminants in stack gases.

Moisture Content of the Raw Gas Before the Gas Cleaning Train

Two sets of tests were performed:

a) The gas was sampled with a stainless steel tube reaching into the middle of the gas producer (shown in Figure 10-28). The stainless steel tube was located 5 cm above the wiper grate. These measurements can therefore be considered a true measure of the water content of the raw gas as a function of temperature and producer gas output. The results are given in Figure 10-50.

It is noted that the moisture content data are reported as grams per m^3 of gas at 25°C, and as percent volume. The latter representation was obtained by assuming that the ideal gas law holds. If we let m be the mass of water condensed from each m^3 (measured at 25°C) of producer gas in an ice bath, then this water would occupy a volume of

$$V_w = \frac{m\ R\ T}{M_v\ p}$$

where $p = 1$ atm and $T > 100°C$.

Each m^3 of producer gas occupies a volume $V_g = \frac{T \cdot 1}{298}$ at temperature T.

Consequently, the water occupies a volume fraction $\varepsilon = \frac{V_w}{V_g + V_w}$. The experimental results confirm the general trend outlined in Section 9.4. The only sources of steam in the gas producer arise from the chemically bound water in the fuel, the natural moisture present in the rice hulls, and the water generated by the water shift reaction. The moisture content of the air has been ignored. Because almost no steam conversion takes place at lower temperatures (< 700°C), a significantly higher water content in the raw gas must be expected as shown.

Moisture content of the raw gas obtained from the design shown in Figure 10-34.

In this test series, the water content of the gas was measured as it left the gas exit, before entering the wet scrubber. It is noted that this data is heavily influenced by the following:

a) The temperature at the gas exit.

b) The temperature at the bottom of the gas producer column.

c) The height of the hydraulic water seal, and water evaporated from it.

The results are given in Table 10-9.

Table 10-9. Water Content of Raw Gas

Gas Flow	Water Content	
$m^3\ h^{-1}$	$g\ m^{-3}$	% volume
3.76	152.6	17.1
4.97	134.6	15.4
5.74	207.5	21.7
7.46	166.2	18.2
8.68	133.0	15.3
9.9	142.2	16.2

In general, one can expect steam contents of 15% - 20% by volume in the hot raw gas. With respect to the "true" steam content of the raw gas as shown in Figure 10-50, it is of interest to note that the steam content decreases with increasing fire zone temperature caused by higher air input rates.

The approximate temperature of the upward moving fire zone as observed in the refractory insulated design (Figure 10-28) is shown in Figure 10-51. The temperatures were measured 1 cm away from the wall and should be viewed as approximate values because thermocouples reaching into a rice hull bed disturb the fuel bed structure. Due to these observed temperatures, the design shown in Figure 10-34 (which does not have a refractory wall) uses a stainless steel tube as inner cylinder.

Moisture Content of the Gas After the Gas Cleaning Train

Producer gas leaving a downdraft gas producer will always be saturated with moisture at temperatures below 100°C. It is therefore of interest to compute the amount of water released when cooling producer gas. The specific humidity, ω, is defined as the ratio of the mass of water vapor to the mass of noncondensible gas. Assuming the ideal gas law holds, we have:

$$\omega = \frac{M_v}{M_g} \frac{P_v}{P_g}$$

where M_v and M_g are the molecular weights of water and producer gas respectively. The molecular weights of the measured producer gas are within the range of 26.8 to 27.9 as shown in Table 10-6. An average of

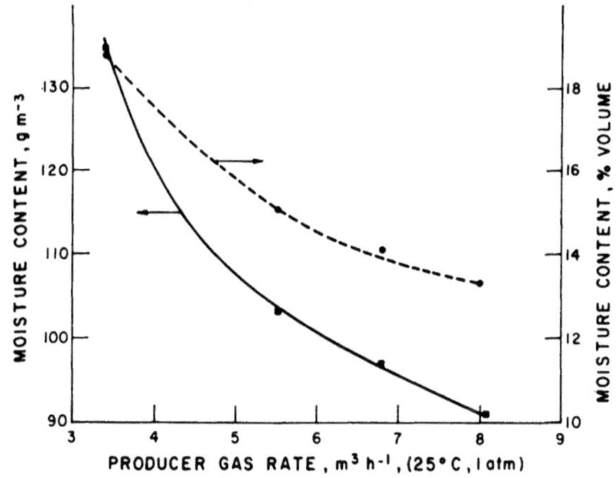

Figure 10-50. Steam content of the raw gas as a function of gas flow rate.

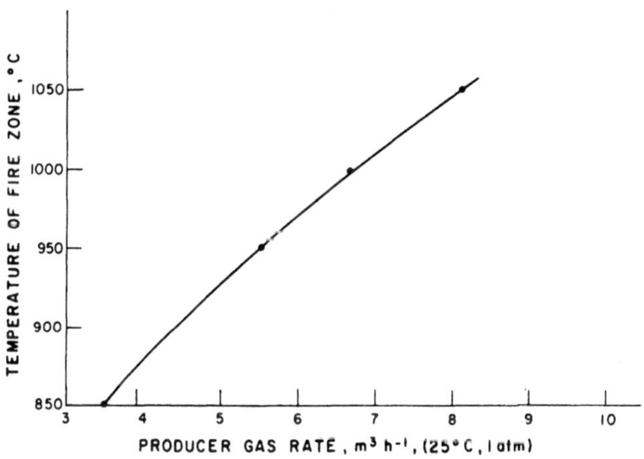

Figure 10-51. Approximate temperature of upward moving fire zone.

27.1 is assumed and therefore $\omega = 0.664 \frac{P_v}{P_g}$.

The relative humidity, ϕ, is defined as the ratio of the actual mass of vapor to the mass of vapor required to produce a saturated mixture.

Therefore $\phi = \frac{p_v}{p_s}$ where p_s is the saturation pressure of the water vapor at the mixture temperature. Consequently $\phi = \frac{\omega \, p_g}{0.664 \cdot p_s}$

Because no vapor pressure curves could be found for producer gas-water mixtures, the curves for air-water mixtures are used. The computational error is probably not large since both air and producer gas consist mainly of nitrogen.

The mass of water vapor carried at $\phi = 1$ and $T = 50°C$ in a producer gas mixture is therefore:

$$\frac{\text{mass of vapor}}{\text{mass of mixture}} = \frac{\omega}{1 + \omega}$$

Since $\phi = \frac{P_v}{P_s}$ it follows that $P_v = P_s = 12350 \text{ N m}^{-2}$ and therefore

$$P_g = (1.01322)(10^5) - 12350 = 88972 \text{ N m}^{-2}$$

Then $\omega = \frac{(0.664)(12350)}{88972} = 0.092$

and $\frac{\text{mass of vapor}}{\text{mass of mixture}} = 0.0842$

When cooling down this mixture to 30°C we have $\frac{\text{mass of vapor}}{\text{mass of mixture}} = 0.029$

Assuming that the mass of gas is preserved during the cooling process, it is concluded that 1 kg of saturated producer gas at 50°C will release 56.9 grams of water when cooled to 30°C. This calculation demonstrates that wet scrubbers should operate at near ambient temperatures. It is noted that the gas leaving a wet scrubber is not only saturated but also contains water mist.

The moisture content of the gas leaving the sieve plate scrubber and dry filter arrangement is given in Table 10-10.

Table 10-10. Moisture Content of Producer Gas after the Gas Cleaning Train

Flow Rate	Moisture Content	
$m^3\ h^{-1}$	$g\ m^{-3}$	% volume (> 100°C)
2.76	71.59	8.9%
4.97	108.13	12.8%
7.46	213.3	22.2
8.68	170.3	18.6
9.9	180.0	19.4

It is noted that the mass of moisture in the gas in Table 10-10 refers to the amount that was condensed out when cooling the gas from the dry gas filter exit temperature down to 13°C - 17°C. The results indicate that at low flow rates the highly absorbent rice hull char can eliminate a significant amount of moisture. For instance, the dry char filter material, which weighed 1037 gram before the test, picked up 577 grams of water at the 4.97 $m^3\ h^{-1}$ flow rate.

Particle Content of Producer Gas

The particle content of the gas was measured downstream from the various gas cleaning arrangements, which are listed as follows:

a) Sieve plate scrubber only with no dry bed filter, operated at a gas flow rate of 7.46 $m^3 h^{-1}$ (data point 1).

b) Sieve plate scrubber followed by a dry bed filter filled with rice hull char (30 cm height). A stainless steel wire mesh demister was placed on top of the rice hull char. Although the steel wire mesh demister does not aid in gas cleaning, it does prevent rice hull char from being entrained in the gas stream. The gas flow was 8.68 $m^3 h^{-1}$ and 9.9 $m^3 h^{-1}$ (data points 2 and 3).

c) Sieve plate scrubber followed by a dry bed filter filled with compacted rice hull char (30 cm height) and the demister on top. The gas flow rate was 4.97 and 3.76 $m^3 h^{-1}$ (data points 4 and 5 respectively).

All data for particle content are given in Figure 10-52.

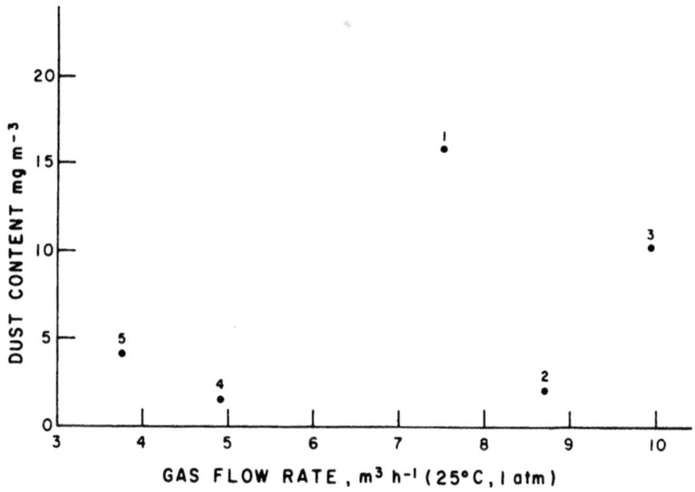

Figure 10-52. Dust content of gas after gas filter train.

The following conclusions can be drawn:

a) A packed bed of rice hull char will further reduce the solid particle content of the gas leaving the wet sieve plate scrubber. This is not obvious if one considers the fact that rice hull char contains many fine particles that could be entrained into the gas stream.

b) The particle load in the gas stream was very low for all tested gas flow rates and was well under the accepted maximum for engine applications.

The appearance of the fiberglass filter papers used for sampling is shown in Figure 10-53. All measurements were done with a precision balance with an accuracy of 10^{-5} grams. Due to humidity in the environment, the mass measured contains an uncertainty factor of $\pm$ 3 mg. This

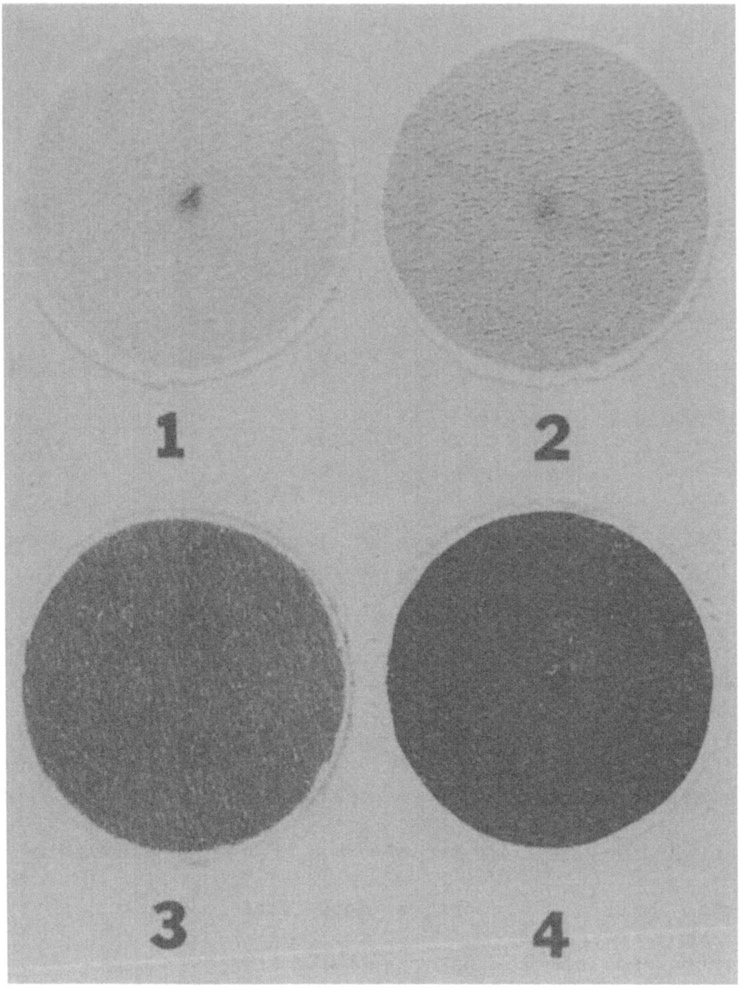

Figure 10-53. Appearance of fiberglass filter papers for total collection of 1) 2.5 mg; 2) 4.3 mg; 3) 17.4 mg; 4) 33.5 mg.

mass represents the amount of moisture a filter paper carries when exposed to ambient air.

Pressure Drop Across the Wet Sieve Plate Scrubber

The pressure drop across the sieve plate scrubber is strongly dependent on the following:

a) The dry plate pressure drop.

b) The height of the water above the sieve plate.

c) The elimination of producer gas flow between the rim of the plate and the wall of the sieve plate scrubber.

The pressure drop across the wet scrubber was kept at 2.5 - 5 cm H_2O at all flow rates. The pressure drop can be easily regulated by adjustment of the water level above the sieve plate (1 - 3 cm). The generated foam height (2 - 8 cm) was sufficient to clean the gas as shown.

Summary

The combination of a wet sieve plate scrubber followed by a dry packed bed filter has proven to be a very successful gas cleaning train. However to avoid large amounts of water being carried out either in mist or vapor form it is suggested the sieve plate scrubber be operated at low temperatures. This may either require very low flow rates for a given cross section or a modest continuous water replacement.

10. LIST OF SYMBOLS

A	total area of sieve plate (m^2)
A_o	hole area of sieve plate (m^2)
C_D	drag coefficient
d	diameter of gas producer, particle diameter, pitch of sieve plate holes.
d_o	diameter of sieve plate holes (mm)
E	volume fraction occupied by the char
g	standard gravitational acceleration, 9.81 m s^{-2}
h	distance traveled during time t* (cm)
M_g	molecular weight of producer gas
M_v	molecular weight of water
m	mass of water condensed out of the gas stream (kg)
m_c	mass of residue (kg)
m_f	mass of rice hulls (kg)
P_g	partial pressure of the gas (N m^{-2})
P_s	saturation pressure of water vapor in air (N m^{-2})
P_v	partial pressure of water vapor (N m^{-2})
R	rice hull consumption rate (kg h^{-1}, wet basis)
Re	Reynolds number
T	temperature (°C)
T_1	fuel bed temperature (°C)
t^*	time needed to reach 99% of the terminal velocity u_t (s)
$u(t)$	particle velocity as a function of time (m s^{-1})
u_t	terminal particle velocity (m s^{-1})
v_s	superficial gas velocity (m s^{-1})

v_a actual gas velocity (m s^{-1})

V_w volume at T = 100°C of collected water vapor (m^3)

V_g volume at T = 100°C of producer gas (m^3)

x_c mass fraction of carbon in rice hull residue (kg)

x_f mass fraction of carbon in rice hulls (kg)

ε volume fraction occupied by water in a gas-water vapor mixture

ε_β void space fraction

Δz height of control volume V (cm)

ω specific humidity

ϕ equivalence ratio and relative humidity

η producer gas viscosity (kg m^{-1} s^{-1})

τ carbon conversion

10. REFERENCES

1. Allen, T. 1974. Particle Size Measurements. Halsted Press, New York.

2. Beagle, E. 1978. Rice husk conversion to energy. FAO Agriculture Services Bulletin #31. Rome.

3. Beagle, E. 1983. Personal communication.

4. Groeneveld, M.J. 1980. The co-current moving bed gasifier. Thesis, Technische Hogeschool Twente, The Netherlands.

5. Calvert, S., et. al. 1972. Wet Scrubber System Study. Volume I: Scrubber Handbook. NTIS, Springfield, VA 22161.

6. Garbarino, G. 1911. Utilizzazione integrale della lolla di riso come combustibile.

7. Harter-Seberick, R. 1937. Motor gas-producers and their fuels, their state of development with special reference to the use of fossil fuels. Fuel 16(1). London.

8. Heiss, F. and J. Coull. 1952. Chemical Engineering Progress 48(3):133-140.

9. Jenkins, B.M. 1980. Downdraft gasification characteristics of major California residue-derived fuels. PhD Thesis in Engineering, University of California, Davis.

10. Kaupp, A. 1982. Technical reports #1 - 14 to the Briggs and Stratton Corporation.

11. Peiyi, Jia, et. al. 1982. Utilization of agricultural wastes for energy conversion and product processing. FAO Bangkok Report RAPA No. 61.

12. Rambush, N.E. 1923. Modern Gas Producers. Van Nostrand Company, New York.

13. Reed, T. 1983. Personal communication. Solar Energy Research Institute, Golden, Colorado.

14. "Buck" Rogers Company. 1983. Personal communication.

15. Reines, R. 1982. Personal communication. Open University at Milton Keynes, Great Britain.

16. Schillar, L. and A.Z. Nauman. 1933. Verein Deutsche Ingenieure Journal 77:318.

17. Schläpfer, P. and J. Tobler. 1937. Theoretische und praktische Untersuchungen über den Betrieb von Motor Fahrzeugen Mit Holzgas. Schweizerische Gesellschaft für das Studium der Motorbrennstoffe, Bern.

18. SERI. 1981. Gengas, the Swedish classic on wood-fueled vehicles. Tipi Workshop Books, Allenspark, Colorado 80510.

19. Skov, N.A. and M.L. Papworth. 1975. The Pegasus Unit. Pegasus Publisher Inc., Olympia, Washington.

20. Treybal, R. 1981. Mass Transfer Operations. McGraw-Hill Publishers, New York.

21. Williams, R.O. and B. Horsfield. 1977. Generation of low-BTU fuel gas from agricultural residue experiments with a laboratory-scale gas producer.

INDEX

	PAGE

GPSR Compliance

The European Union's (EU) General Product Safety Regulation (GPSR) is a set of rules that requires consumer products to be safe and our obligations to ensure this.

If you have any concerns about our products, you can contact us on ProductSafety@springernature.com

In case Publisher is established outside the EU, the EU authorized representative is:

Springer Nature Customer Service Center GmbH
Europaplatz 3
69115 Heidelberg, Germany

Batch number: 10578397

Printed by Printforce, the Netherlands